Green and Sustainable Medicinal Chemistry
Methods, Tools and Strategies for the 21st Century Pharmaceutical Industry

RSC Green Chemistry

Titles in the Series:

1: The Future of Glycerol: New Uses of a Versatile Raw Material
2: Alternative Solvents for Green Chemistry
3: Eco-Friendly Synthesis of Fine Chemicals
4: Sustainable Solutions for Modern Economies
5: Chemical Reactions and Processes under Flow Conditions
6: Radical Reactions in Aqueous Media
7: Aqueous Microwave Chemistry
8: The Future of Glycerol: 2nd Edition
9: Transportation Biofuels: Novel Pathways for the Production of Ethanol, Biogas and Biodiesel
10: Alternatives to Conventional Food Processing
11: Green Trends in Insect Control
12: A Handbook of Applied Biopolymer Technology: Synthesis, Degradation and Applications
13: Challenges in Green Analytical Chemistry
14: Advanced Oil Crop Biorefineries
15: Enantioselective Homogeneous Supported Catalysis
16: Natural Polymers Volume 1: Composites
17: Natural Polymers Volume 2: Nanocomposites
18: Integrated Forest Biorefineries
19: Sustainable Preparation of Metal Nanoparticles: Methods and Applications
20: Alternative Solvents for Green Chemistry: 2nd Edition
21: Natural Product Extraction: Principles and Applications
22: Element Recovery and Sustainability
23: Green Materials for Sustainable Water Remediation and Treatment
24: The Economic Utilisation of Food Co-Products
25: Biomass for Sustainable Applications: Pollution Remediation and Energy
26: From C–H to C–C Bonds: Cross-Dehydrogenative-Coupling
27: Renewable Resources for Biorefineries
28: Transition Metal Catalysis in Aerobic Alcohol Oxidation
29: Green Materials from Plant Oils
30: Polyhydroxyalkanoates (PHAs) Based Blends, Composites and Nanocomposites

31: Ball Milling Towards Green Synthesis: Applications, Projects, Challenges
32: Porous Carbon Materials from Sustainable Precursors
33: Heterogeneous Catalysis for Today's Challenges: Synthesis, Characterization and Applications
34: Chemical Biotechnology and Bioengineering
35: Microwave-Assisted Polymerization
36: Ionic Liquids in the Biorefinery Concept: Challenges and Perspectives
37: Starch-based Blends, Composites and Nanocomposites
38: Sustainable Catalysis: With Non-endangered Metals, Part 1
39: Sustainable Catalysis: With Non-endangered Metals, Part 2
40: Sustainable Catalysis: Without Metals or Other Endangered Elements, Part 1
41: Sustainable Catalysis: Without Metals or Other Endangered Elements, Part 2
42: Green Photo-active Nanomaterials: Sustainable Energy and Environmental Remediation
43: Commercializing Biobased Products: Opportunities, Challenges, Benefits, and Risks
44: Biomass Sugars for Non-Fuel Applications
45: White Biotechnology for Sustainable Chemistry
46: Green and Sustainable Medicinal Chemistry: Methods, Tools and Strategies for the 21st Century Pharmaceutical Industry

How to obtain future titles on publication:
A standing order plan is available for this series. A standing order will bring delivery of each new volume immediately on publication.

For further information please contact:
Book Sales Department, Royal Society of Chemistry, Thomas Graham House, Science Park, Milton Road, Cambridge, CB4 0WF, UK
Telephone: +44 (0)1223 420066, Fax: +44 (0)1223 420247
Email: booksales@rsc.org
Visit our website at www.rsc.org/books

Green and Sustainable Medicinal Chemistry
Methods, Tools and Strategies for the 21st Century Pharmaceutical Industry

Edited by

Louise Summerton
University of York, UK
Email: louise.summerton@york.ac.uk

Helen F. Sneddon
GlaxoSmithKline, Stevenage, UK
Email: helen.f.sneddon@gsk.com

Leonie C. Jones
University of York, UK
Email: leonie.jones@york.ac.uk

James H. Clark
University of York, UK
Email: james.clark@york.ac.uk

ROYAL SOCIETY OF CHEMISTRY

THE QUEEN'S AWARDS FOR ENTERPRISE: INTERNATIONAL TRADE 2013

RSC Green Chemistry No. 46

Print ISBN: 978-1-78262-467-7
PDF eISBN: 978-1-78262-594-0
EPUB eISBN: 978-1-78262-747-0
ISSN: 1757-7039

A catalogue record for this book is available from the British Library

Published by The Royal Society of Chemistry,
Thomas Graham House, Science Park, Milton Road,
Cambridge CB4 0WF, UK

Registered Charity Number 207890

For further information see our web site at www.rsc.org

Printed in the United Kingdom by CPI Group (UK) Ltd, Croydon, CR0 4YY, UK

FOREWORD

An Introduction to CHEM21 Chemical Manufacturing Methods for the 21st Century Pharmaceutical Industries

MURRAY J. B. BROWN

GSK Medicines Research Centre, Gunnels Wood Road, Stevenage, Hertfordshire SG1 2NY, UK
Email: Murray.J.Brown@gsk.com

1 Sustainability in Pharmaceutical Manufacturing

The sustainable use of resources is essential in all areas of business, and pharmaceuticals is no exception. There are regulatory and legislative pressures on health care providers as well as individual manufacturers. For example, the National Health Service (NHS) in the UK has a Sustainable Development Strategy[1] with an emphasis on reducing the environmental impact of the health and care system to improve economic, social and environmental sustainability. Pharmaceuticals have been identified as a "carbon hotspot" with 21% of NHS greenhouse gas emissions attributable to pharmaceuticals.[2] Similarly, the Swedish Medical Products Agency has an emphasis on sustainability[3] and the US Environmental Protection Agency sponsors the prestigious Presidential Green Chemistry Challenge with at least 11 previous winners having pharmaceutical applications.[4] Further drivers for sustainability, such as REACH legislation, risks associated with

RSC Green Chemistry No. 46
Green and Sustainable Medicinal Chemistry: Methods, Tools and Strategies for the 21st Century Pharmaceutical Industry
Edited by Louise Summerton, Helen F. Sneddon, Leonie C. Jones and James H. Clark

Published by the Royal Society of Chemistry, www.rsc.org

non-renewable fossil feedstocks and elemental sustainability, are described in Chapter 1 (Green and Sustainable Chemistry: An Introduction by James Clark). Sustainability also makes good business sense and pharmaceutical companies have sustainability objectives as part of their corporate responsibility goals.[5] For example, GSK is committed to being carbon neutral by 2050 and to halve operational waste by 2020.[6]

Medicines present unique challenges and opportunities for sustainable manufacture compared to larger volume bulk commodities (see Chapter 8: From Discovery to Manufacturing: Some Sustainability Challenges Presented by the Requirements of Medicine Development by John Hayler). The active pharmaceutical ingredient (API) often has high molecular complexity and stringent specifications are applied to the amounts and identities of impurities allowed requiring very high levels of chemoselectivity in their manufacturing routes. Medicines are often produced with very low mass efficiencies such that less than 1 kg of API is produced from 100 kg of input materials. In part, this is due to the requirement to use bulky protecting groups to ensure appropriate selectivity and large quantities of solvent to facilitate downstream processing steps, such as crystallisation, to ensure purity. Additionally, pharmaceutical syntheses often use stoichiometric rather than catalytic reagents. The resulting cumbersome routes require large, often dedicated, facilities and produce long lead times such that the synthesis may have to start months or even years before a pill reaches a patient. The generation of large quantities of waste apparent from these low mass efficiencies resulting in consequent material and waste treatment costs are the most obvious sustainability issues. A further sustainability challenge is that where more efficient catalytic processes are used they are usually reliant on precious metals such as platinum and palladium. Demand for these metals from consumer industries beyond pharmaceuticals and geo-political concerns around security of supply raise the possibility that they will become too scarce to be economically viable for use in making medicines (see Chapter 5: The Importance of Elemental Sustainability and Critical Element Recovery for the Pharmaceutical Industry by Andrew Hunt). To ensure sustainable delivery of drugs to patients and ensure pharmaceutical manufacturing competiveness, it is essential to develop 'green' alternatives to existing reaction steps, new methodologies to shorten routes and lead times and innovative technologies to intensify processes and reduce manufacturing footprint.

2 IMI's Call to Action

In July 2011, the Innovative Medicines Initiative (IMI) published a call entitled "Sustainable Chemistry—Delivering Medicines For The 21st Century".[7] IMI is a partnership between the European Union (represented by the European Commission) and the European pharmaceutical industry (represented by EFPIA, the European Federation of Pharmaceutical Industries and Associations), including the big pharma global corporations as well as small and medium healthcare enterprises. The EFPIA member companies recognised that both society and the pharmaceutical industry derive substantial benefit from the invention of medicines that allow patients to live longer, healthier, and more productive lives. In addition, pharmaceutical companies are committed to finding novel ways to bring additional value, such as developing key medicines with minimum impact on the environment. Thus, the pharmaceutical industry is embracing the concept of Sustainable Chemistry in its broadest forms as the ideal framework to develop synthetic methodologies that minimise environmental impact. Backing up that sentiment, the EFPIA members were prepared to commit over € 10 million as in-kind contribution, a sum to be matched by the European Commission to fund public research in the area.

The key reactions used within medicinal and process chemistry for the pharmaceutical industry are well documented and the key sustainability challenges have previously been described, for example by the ACS Green Chemistry Institute Pharmaceutical Roundtable (ACS GCI PR). For the purposes of the IMI call, the EFPIA members focussed on the following areas of outstanding need:

- Amide formation to improve atom economy
- Aliphatic and aromatic selective C–H activation by catalysis for C–C, C–O, and C–N formation
- Asymmetric synthesis for quaternary stereo centres
- Chiral amine synthesis (N-centered chemistry)
- Alcohol activation for nucleophilic substitution
- Green fluorination (selectivity/reagents)

It was further suggested that the successful applicants should update the seminal review of Constable *et al.*[8] produced by the ACS GCI PR to develop a detailed understanding of requirements beyond 2020 and highlight changes in landscape that could be foreseen.

Key enabling technologies for sustainable chemistry have also been identified by various stakeholders. The IMI call highlighted areas of promise that have yet to penetrate significantly into pharmaceutical manufacturing:

- Development and use of novel organic and organometallic catalysis, particularly using base metals
- Intensification of processes, for instance using continuous (flow) chemistry to improve space–time yields from manufacturing plants
- Development and industrialisation of biocatalysts to improve process metrics (*e.g.* by obviating the need for protecting groups) or enable novel routes to APIs or intermediates
- Synthetic Biology as a technology to allow the cascading of biocatalytic reactions in sequence, exploiting developments in high throughput molecular biology to deliver organisms capable of producing high value molecules

Lastly, but by no means least, the EFPIA members realised that educating and motivating medicinal and development chemists to adopt new technologies and methodologies is an essential component. Medicinal chemists generally do not have a focus on sustainability as they have to achieve the complex and expensive task of finding drug candidates conforming to increasingly stringent criteria around pharmacodynamics, pharmacokinetics, toxicity profile, *etc.*, and their environmental impact is generally minimal as they work at small scale. However, the routes defined in the early discovery phases rapidly become embedded during clinical phases as impurity profiles are defined and toxicity trials are performed. There are significant time and financial pressures to continue with an initial route, in good part driven by high attrition in the clinic meaning investment to change a route is often not rewarded. Thus, initial poorly sustainable route choices early in development can get locked in to the final manufacturing process unless conscious decisions are made to adopt more sustainable alternatives as early as possible. The next generation of medicinal and process chemists needs a far greater awareness of sustainability issues and technologies and methodologies to address them. A key component of the IMI call was to provide education and training materials to help effect that change.

After the call was published, a number of high-quality expressions of interest were received and the best were invited to progress to a second stage to form Full Consortia with the corresponding EFPIA participants and to prepare and submit Full Project Proposals to IMI JU. Finally, at the end of 2012 CHEM21 was launched to develop chemical manufacturing methods for the 21st century pharmaceutical industries.

3 The CHEM21 Project

The CHEM21 Consortium is composed of 10 University research groups, 5 Small and Medium Enterprises (SMEs) and 6 EFPIA member companies (Table 1).

Funding to the Universities, Research Organisations, Public bodies and SMEs is provided *via* the IMI under grant agreement n°115360, which matches the in kind contribution of the EFPIA and amounts to a total budget of € 26.4M. The partners span the breadth of Europe (see Figure 1) and bring a wealth of complementary skills and experience.

The stated aim of CHEM21 is to develop a broad based portfolio of sustainable technologies for green chemical intermediate manufacture aimed at the pharmaceutical industry, and to train and educate scientists in these technologies and methodologies. The project was designed with an initial 'base-lining' work package to survey the landscape and produce 'Vision 2020' as a roadmap for developing the appropriate sustainable technologies for future manufacturing. The technologies being developed were divided into three work packages based on chemical catalysis and synthetic methods, biocatalysis and synthetic biology. A further work package was designed to develop and deliver training and education packages for a range of

Table 1 Consortium Partners in CHEM21 project.

Partner	Location
Universities, Research Organisations and Public bodies	
University of Manchester	Manchester, UK
Leibniz Institute for Catalysis (LIKAT)	Rostock, Germany
Stichting VU-VUMC	Amsterdam, Netherlands
Technische Universität Graz	Graz, Austria
Universität Graz	Graz, Austria
Universität Stuttgart (Institute for Technical Biochemistry)	Stuttgart, Germany
Universiteit Antwerpen	Antwerp, Belgium
University of Durham	Durham, UK
University of Leeds	Leeds, UK
University of York	York, UK
SMEs	
Austrian Center of Industrial Biotechnology (ACIB GmbH)	Graz, Austria
CatScI Ltd	Wentloog, Cardiff, UK
Charnwood Technical Consulting Ltd	Quorn, UK
Evolva Biotech A/S	Copenhagen, Denmark
Reaxa Limited	Leeds, UK
EFPIA	
GlaxoSmithKline Research and Development Ltd	Brentford, UK
Bayer Pharma AG	Berlin, Germany
Janssen Pharmaceutica NV	Beerse, Belgium
Orion Corporation	Espoo, Finland
Pfizer Limited	Sandwich, UK
Sanofi Chimie	Gentilly, France

audiences. The University of Manchester stand as the Managing Entity, providing financial management, and GSK act as Coordinator to engage with IMI. Together they lead the management group, which provides governance for the project to ensure nothing impedes the scientific progress of the key work packages. Thus, the initial work package defines the landscape, the three scientific work packages measure, analyse and implement solutions to generate an improved toolbox of sustainable manufacturing technologies and the final work packages provide for control to ensure embedding of the advances in a DMAIC style data-driven improvement cycle (Figure 2).

4 CHEM21 Work Packages

4.1 Work Package 1: Base-lining and Prospecting the Manufacturing Landscape

The base-lining work package is now complete to deliver 'Vision2020' and the findings are in the process of being published.[9–12] The problem statement of the original call and the initially proposed technological solutions by the consortium were confirmed as still highly relevant, which was

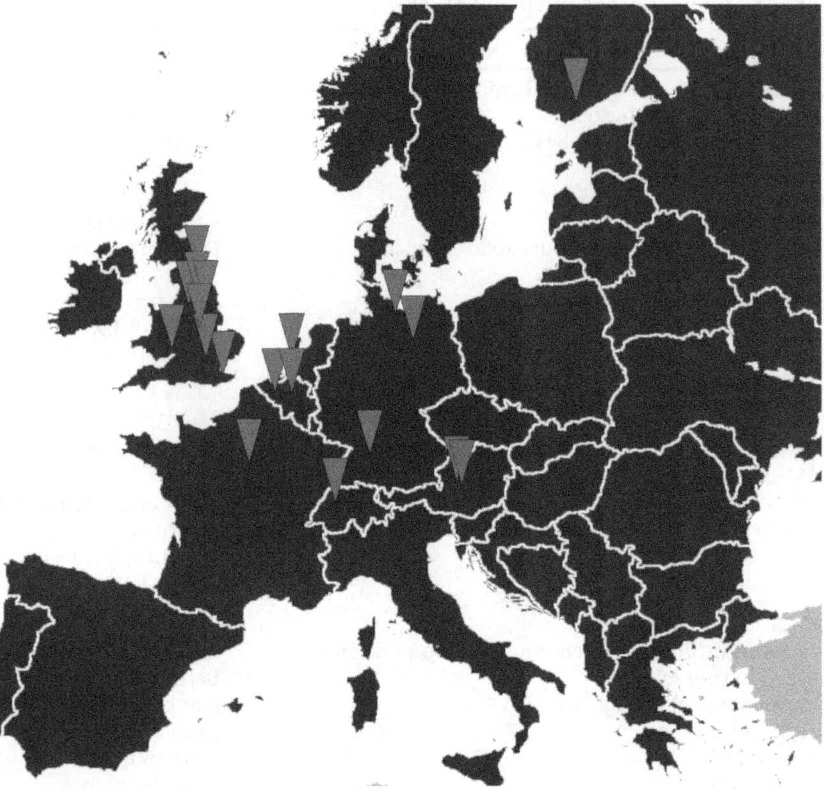

Figure 1 Map of locations of CHEM21 partners.

encouraging but also an indication of the intractability of many of the issues and the slow pace of change in the pharmaceutical industry. The description of work and the operating ethos of CHEM21 also matched external stakeholder expectations. In addition to developing novel reactions, a clear need and stakeholder desire to drive sustainability through metrics, training and education, improved solvent selection, and a move towards greater use of biotechnology solutions was highlighted. Although the work plan for CHEM21 does cover some of the reaction classes identified as desirable for increasing diversity for medicinal chemistry, a number of additional areas were identified that could not be covered. It would be unrealistic for even a project of the size of CHEM21 to cover all possible medicinal and process chemistry needs, and indeed it is unrealistic for a complete overlap of medicinal chemistry and process chemistry space. Some of the issues are highlighted in Chapter 2: Tools For Facilitating More Sustainable Medicinal Chemistry and Chapter 9: Medicinal Chemistry: How "Green" Is Our Synthetic Tool Box?, and Biocatalysis for Medicinal Chemistry is covered in Chapter 15. The base-lining effort also emphasised that ways of working and effective collaboration will be the key to success of CHEM21, for example

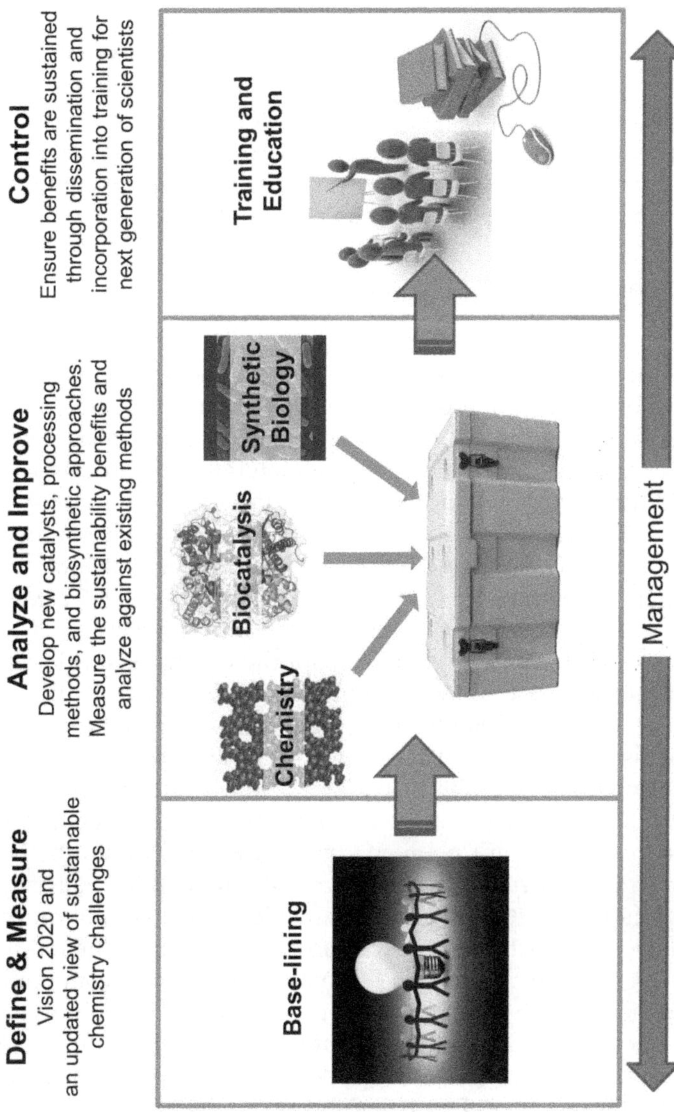

Figure 2 Structure and interrelationship of CHEM21 project work packages (figure kindly provided by Professor Nick Turner).

the selection of suitable target molecules, close association of the EFPIA members with the academic research programs and willingness for EFPIA members to adopt any new technologies developed when business drivers make sense to do so. Subsequently, CHEM21 partners have identified a number of APIs appearing on the WHO Essential Medicine's List that are suitable for the technologies being developed in addition to EPFIA partners' proprietary drugs.

Finally, the working group highlighted a need to focus on understanding industry needs and the reasons and barriers for previous technology/market failures in rapid adoption of greener methodology. This challenge is not to be underestimated in such a conservative industry and emphasises the importance of the training and education of current and future medicinal and process chemists in the use of sustainable technologies.

4.2 Work Package 2: Chemical Technologies

The overarching aim of the second work package is to develop an improved sustainable chemistry toolbox for Medicinal and Process Chemists to use to develop greener routes for the manufacture of pharmaceutical APIs. The intent is to do this *via* a focus on a relatively small number of simple and broadly applicable reactions using starting materials that are atom-efficient and have reduced levels of benign waste, requiring less solvent that is more easily re-used. For catalytic methods there is a focus on using abundant and less toxic base metals, such as iron, copper, and nickel (Chapter 16: Base Metals in Catalysis: From Zero to Hero, by Bert Maes *et al.*). Where precious metals are necessary, more sustainable use and recovery is a priority (Chapter 11: Pd-catalysed Cross-couplings for the Pharmaceutical Sector and a Move to Cutting-edge C–H Bond Functionalization: is Palladium Simply Too Precious? by Ian Fairlamb). Continuous flow methods for reaction, work-up and isolation are important means to achieve process intensification and economic viability leading to more sustainable processes (Chapter 12: The Growing Impact of Continuous Flow Methods on the Twelve Principles of Green Chemistry, by John Blacker). Fluorine is a very important element in drug design and presents some unique sustainability challenges (Chapter 17: 'Green' and Sustainable Halogenation Processes, by Graham Pattison) and CHEM21 is evaluating the use of elemental fluorine for greener fluorination. Amide bond formation is the most commonly used synthetic reaction in both medicinal and process chemistry. While established methods are reliable and effective they are not particularly green, so work package 2 is evaluating catalytic amide bond formation (see Chapter 13: Green Catalytic Direct Amide Bond Formation, by Andrew Whiting, for an overview of the area). Finally, the work package will scale-up successful reactions and measure the green performance using the sophisticated metrics developed within the CHEM21 project (Chapter 4: Beyond Mass-based Metrics: Evaluating the Greenness of Your Reaction, by Louise Summerton) to demonstrate significant improvement over existing methodologies. An emphasis is to

- The potential for sustainable industrial production
- Potentially improved community health
- Supports rural development
- Increased industrial competitiveness through innovative eco-efficient products
- Potential for transfer to other regions of the world, including the transfer of appropriate technologies discovered and proven in the EU

Vijayendran recently estimated that by 2025 over 15% of the $3 trillion global chemical market will be derived from bio-sources.[11] Active pharmaceutical ingredients (APIs), polymers, cosmetics, lubricants and solvents have also been estimated to be the most important sub-segments of the chemical sector by the *ad hoc* advisory group for bio-based products.[10] APIs in particular, with 33.7% of global chemical sales, are expected to be the chemical segment with the highest percentage sales of products produced using biotechnological processes. As we begin to move away from petrochemicals, the use of biomass as a chemical feedstock will become increasingly important.[12]

1.3 Biomass as a Chemical Feedstock

Biomass is generally understood to mean large-volume, low-value bioresources that can be used as feedstocks for making chemicals, fuels and materials. To distinguish biomass from fossil resources, such as coal and petroleum (themselves ancient biomass), it is prudent to limit ourselves to resources that are less than 100–200 years old (resources that have a similar lifecycle to man). In this way, we can consider biomass to include:

- Forestry residues
- Short rotation trees
- Agricultural residues including straws
- Food processing wastes including shells, stones, peels
- Grasses and other land grown biomass not used for food
- Marine residues
- Macroalgae (seaweed), microalgae and other water-grown biomass not used for food
- Other food wastes

The total amount of this available biomass is not precisely known but has previously been estimated to be 50 billion tonnes per year, including 1.3 billion tonnes per year of food waste.[13,14]

We can classify biomass into 3 main categories:[12]

- Carbohydrate (starch, cellulose and hemicellulose) including lignin from lignocellulosic biomass
- Triglycerides (soybean, palm, rapeseed, sunflower oil)
- Mixed organic residues

Lignocellulosic biomass consists of dry plant matter containing cellulose, hemicellulose and lignin. It can be sourced from a variety of dedicated crops, such as miscanthus, willow or poplar. Alternatively, feedstocks can be obtained from wastes such as rice or wheat straw, forestry residues and paper pulp from the paper industry. Food waste is another feedstock rich in functionalized molecules. Although it is biodegradable, it should be valorized as a raw material for renewable chemicals, materials and biofuels, leading us towards waste minimization and reduced dependence on fossil resources. The utilization of waste materials has the key advantages that it avoids competition for agricultural land that might be used for food production whilst generating value from residues that might otherwise go to waste.[15] For these reasons, waste valorization is seen as an increasingly important source of both chemicals and energy.

In addition to the extractable functional molecules found in biomass, we can also make additional useful functional molecules or "platform molecules", such as succinic acid, lactic acid and levoglucosenone, by biochemical or thermochemical processing of the bulk cellulosic components of many types of biomass. A biorefinery is an analogue to the current petro-refinery in the sense that it produces energy and chemicals. The major differences lies in the raw material it will use, ranging from biomass to waste (Figure 1.2).

Biorefineries can be considered to belong to three types. Type 1 biorefineries focus on the conversion of one feedstock, using one process and targeting one product. A biodiesel production plant would be a good example: rapeseed or sunflower is used for oil extraction, which is subsequently transesterified to produce fatty acid methyl esters or biodiesel using methanol and a catalyst.

Type 2 biorefineries differ from the first type by the number of products. A typical example is the production of starch, ethanol and lactic acid together with high fructose syrup, corn syrup, corn oil and corn meal from corn wet mill operations. A more recent example that has been suggested is the use of citrus waste, such as orange peel (Figure 1.3).[16]

Type 3 biorefineries allow for a wider range of technologies to be combined. They also allow for a higher number of products generating two or more bio-based products and the residue is used to produce energy (either fuel, power and/or heat). Examples include whole crop biorefineries, which make use of several agricultural by-products originating from the same crop. Type 3 biorefineries are typically the ones targeting the production of chemicals and fuels.

As the biorefinery concept is developed, it is imperative that the use of clean technology is applied, ensuring its output(s) are truly sustainable. The IEA Bioenergy Task 42 defines biorefining as "the sustainable processing of biomass into a spectrum of bio-based products (food, feed, chemicals and/or materials) and bioenergy (biofuels, power and/or heat).[17] In the future, various biorefineries will emerge commercially, taking advantage of flexible technology, helping the concept of a biorefinery to process locally available

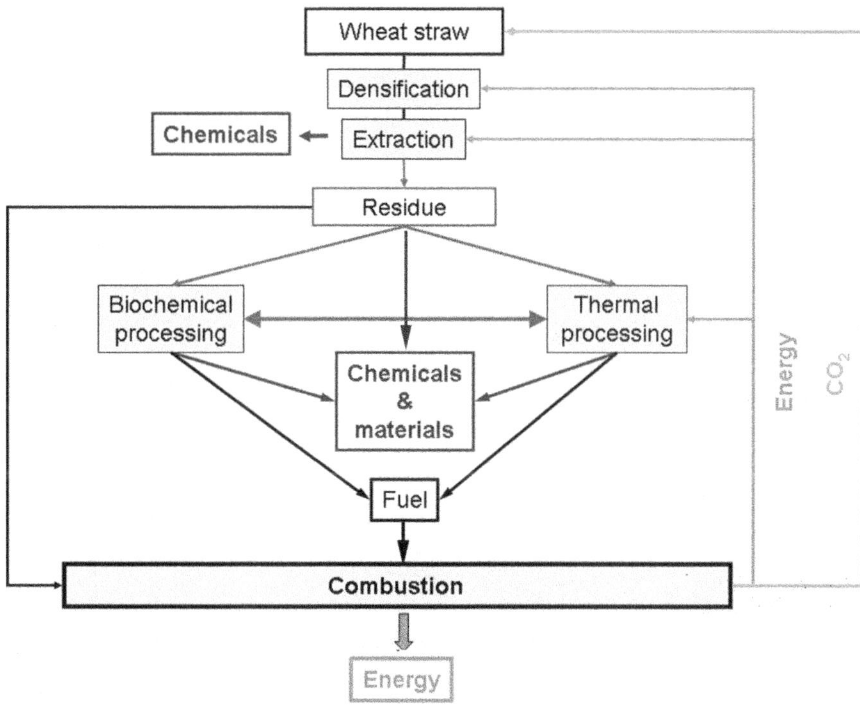

Figure 1.2 Proposed scheme for an integrated biorefinery.
Reproduced from V. L. Budarin, P. S. Shuttleworth, J. R. Dodson, A. J. Hunt, B. Lanigan and R. Marriott *et al.*, *Energy Environ. Sci.*, 2011, **4**, 471 with permission from The Royal Society of Chemistry.[26]

biomass in an integrated fuel-chemical-material-power cycle, improving the quality of life of the local population and lowering the environmental impact governed by the three dimensions of sustainability: environmental protection, social progress and economic development.[12]

Green Chemistry is helping to drive the renewables revolution, pointing the way towards the substitution of fossil feedstocks and towards a more circular economy approach in resource utilization. The three fundamental stages in the product lifecycle are no longer enough—we now need to add a stage that returns the resources from the spent article to useful production. This could in principle be part of the natural carbon cycle for organic materials so that we only need ensure the articles are collected (better infrastructure) and that they are (quickly) biodegradable. The limitation is that nature tends to transfer most of its carbon in the form of carbon dioxide and other simple molecules, which then require effort to build up (by performing chemical reactions *etc.* that consume resources and generate their own waste). In the case of inorganic resources, we cannot rely on any sort of natural cycle for most elements. Our current linear approach of extracting ores, processing them to make metals, using those metals in the manufacture of complex articles, and then disposing of them in landfill sites

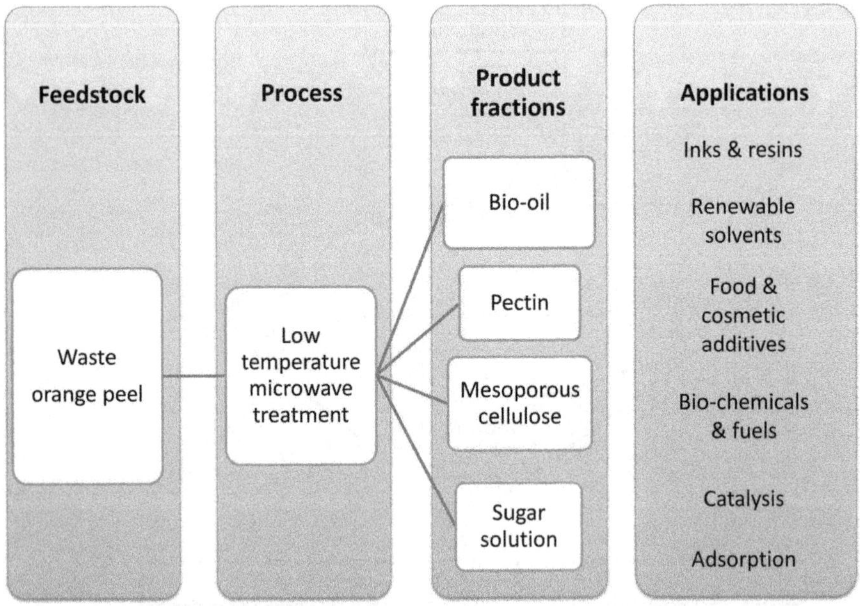

Figure 1.3 Overview of citrus waste valorization utilizing low temperature microwave treatments.

cannot return the resources to us in any useful way. Rather we need to build up our own inorganic resource closed loop systems whereby the resources, typically metals are recovered from the original articles in which they are used, and in a form that can be easily used for the same or different application. This will require fundamental changes in the design of articles enabling easy disassembly at a resource level—sometimes referred to as "benign by design".

1.4 Major Initiatives Worldwide

As has been discussed, increasing legislation, limited resources, and changes in scientific and public opinion mean that there is a growing need for industries and academia to work together towards greener and more sustainable practices. It is vital that we equip the next generation of scientists with the knowledge and skills to do so.

From the humble beginnings of a few research groups working in areas such as the replacement of commonly used hazardous reagents like $AlCl_3$ in the 1980s, through the Green Chemistry movement started by the US EPA in the 1990s, there are now many green and sustainable chemistry initiatives worldwide. These range from large centres working in several areas to educational programs and networks. Green Chemistry is finding its way into various stages of education and in various countries: probably the most recognizable educational programs are the Masters courses now running in countries including the UK, France, Spain, Canada, Greece, India and Bulgaria.[18] Green Chemistry is probably less well developed at undergraduate levels, but in the US, it's been estimated that 13 universities offer Green Chemistry courses with Berkeley being especially active including developing online activities.[11] A number of books and other resources offer Green Chemistry practicals, mostly for undergraduate courses.[19]

Green Chemistry centres (mostly university-based with more than one senior academic based in the Centre, and with a range of activities going beyond research to include for example, education and networking) are becoming widespread with several in the USA (including UC Berkeley and UMass Boston) and others in Australia (Center for Green Chemistry at Monash), Korea (including the Korean Research Institute of Chemical Technology), Mexico (UANL, Monterrey), India (University of Delhi), Canada (Green Center Canada) and UK (University of York).[20]

The Green Chemistry Network (GCN) has recently initiated a network of green and sustainable chemistry centres (G2C2) in an attempt to improve communication between existing centres and to provide a guide for emerging centres, such as in Brazil and South Africa.[21] The first meeting of international centres took place in Delhi in December 2013.[22] In addition, the Green Chemistry Institute is highly active in promoting several important Green Chemistry initiatives, such as the greening of pharmaceutical processes.[23,24] Collaborations between academia and industry are also key in developing cutting-edge technologies and projects, such as the IMI 'CHEM21 (Chemical manufacturing methods for the 21st century pharmaceutical industries)' project.[25] The aim of this project is to work towards developing greener methodologies by developing sustainable biological and chemical alternatives to finite materials, such as precious metals. Many of the themes of the project are covered in the chapters of this book.

1.5 Summary

- Green Chemistry, Green Engineering and Sustainable Design must be considered when moving towards processes that are more environmentally appropriate.
- Changing public opinion, legislation and availability of resources are all driving change.
- The implementation of REACH and development of 'SIN-Lists' of chemicals that aim to restrict the use of many chemicals has wide ranging implication for the manufacturing industries.
- The limited availability of many resources means we must develop closed-loop systems and move towards a circular economy.
- The biorefinery concept allows the conversion of biomass into useful chemicals and energy, reducing our dependence on fossil resources.
- There are many initiatives worldwide that seek to aid innovation in the area of green and sustainable chemistry and educate the next generation of scientists.

Acknowledgements

The research leading to these results has received funding from the Innovative Medicines Initiative Joint Undertaking under grant agreement No 115360, resources of which are composed of financial contributions from the European Union's Seventh Framework Programme (FP7/2007-2013) and EFPIA companies' in kind contribution. IMI www.imi.europa.eu

The views expressed in regards to education and training materials represent the aspiration of the CHEM21 consortium, although may not always be the view of each individual organization.

References

1. P. T. Anastas and J. C. Warner, *Green Chemistry: Theory and Practice*, Oxford University Press, New York, 1998, vol. 1, p. 30.
2. P. T. Anastas and J. B. Zimmerman, *Environ. Sci. Technol.*, 2003, **37**, 94A.
3. http://echa.europa.eu/web/guest/regulations/reach/understanding-reach (last accessed July 2015).
4. M. Lancaster, *Handbook of Green Chemistry and Technology*, ed. J. H. Clark and D. J. Macquarrie, Blackwell, Oxford, 2007, vol. 2, p. 10.
5. http://www.chemsec.org/what-we-do/sin-list (last accessed July 2015).
6. http://echa.europa.eu/web/guest/candidate-list-table (last accessed July 2015).
7. https://www.gov.uk/rohs-compliance-and-guidance (last accessed July 2015).
8. A. J. Hunt and J. H. Clark, *Element Recovery and Sustainability*, ed. A. J. Hunt and J. H. Clark, RSC Publishing, Cambridge, 2013.
9. J. R. Dodson, A. J. Hunt, H. L. Parker, Y. Yang and J. H. Clark, *Chem. Eng. Process.*, 2012, **51**, 69.
10. http://ec.europa.eu/enterprise/sectors/biotechnology/files/docs/bio_based_from_promise_to_market_en.pdf (last accessed July 2015).
11. http://www.nova-institut.de/pdf/11-01%20Biobased%20Chemicals%20White%20Paper%20Asia.pdf (last accessed July 2015).
12. L. A. Pfaltzgraff and J. H. Clark, *Green Chemistry, Biorefineries and Second Generation Strategies for Re-use of Waste: An Overview*, ed. K. W. Waldron, Woodhead Publishers, Cambridge, vol. 1, 2014, p. 3.
13. C. Okkerse and H. van Bekkum, *Green Chem.*, 1999, **1**, 107.
14. http://www.fao.org/docrep/014/mb060e/mb060e00.htm (last accessed July 2015).
15. F. Cherubini, *Energy Convers. Manage.*, 2010, **51**, 1412.
16. L. A. Pfaltzgraff, M. De bruyn, E. C. Cooper, V. Budarin and J. H. Clark, *Green Chem.*, 2013, **15**, 307.
17. http://www.ieabioenergy.com/wp-content/uploads/2013/10/IEA-Bioenergy-2009-Annual-Report.pdf (last accessed July 2015).

18. http://advancinggreenchemistry.org/benchmarking/education/green-chemistry-is-emerging-in-academic-institutions-all-over-the-world/ (last accessed July 2015).
19. *Green Organic Chemistry in Lecture and Laboratory*, ed. A. P. Dicks, CRC Press, Boca Raton, 2012, vol. 1.
20. http://www.idpmultimedia.com.au/greenchemweb/research_networks.html (last accessed July 2015).
21. http://www.greenchemistrynetwork.org (last accessed July 2015).
22. https://g2c2.greenchemistrynetwork.org/ (last accessed July 2015).
23. http://www.acs.org/content/acs/en/greenchemistry.html (last accessed July 2015).
24. http://www.acs.org/content/acs/en/greenchemistry/industry-business/pharmaceutical.html (last accessed July 2015).
25. http://www.chem21.eu/ (last accessed July 2015).
26. V. L. Budarin, P. S. Shuttleworth, J. R. Dodson, A. J. Hunt, B. Lanigan, R. Marriott *et al.*, *Energy Environ. Sci.*, 2011, **4**, 471.

CHAPTER 2

Tools for Facilitating More Sustainable Medicinal Chemistry

H. F. SNEDDON

GSK, Medicines Research Centre, Green Chemistry, Gunnels Wood Road, Stevenage, Herts., SG1 2NY, UK
Email: helen.f.sneddon@gsk.com

2.1 Introduction

Today's medicinal chemists face an array of different pressures.[1] Just as medicinal chemists need a working knowledge of the biology of their target, screening technologies, drug metabolism and pharmacokinetics (DMPK), and the challenges of drug development, an understanding of the principles of green chemistry is also desirable. Detailed metrics on routes and life cycle analysis are entirely appropriate for a route to an active pharmaceutical ingredient (API) that may ultimately be manufactured on a multi-tonne scale. However, such approaches are generally too time-consuming to be used on a day-to-day basis in medicinal chemistry, where multiple routes to multiple different compounds are being considered. Instead simple tools, designed to be quick and easy to use, can help guide medicinal chemists towards making more sustainable choices, with the goal of both minimizing any adverse environmental impact of drug discovery operations, and hopefully minimising major route changes on candidates passed on to development. In addition, these tools also have the potential to promote engagement with

RSC Green Chemistry No. 46
Green and Sustainable Medicinal Chemistry: Methods, Tools and Strategies for the 21st Century Pharmaceutical Industry
Edited by Louise Summerton, Helen F. Sneddon, Leonie C. Jones and James H. Clark

Published by the Royal Society of Chemistry, www.rsc.org

other aspects of green chemistry and wider sustainability initiatives. This chapter seeks to offer an overview of tools available for solvent and reagent selection.

2.2 Solvent Selection Guides

When tackling sustainability in medicinal chemistry, it is logical to focus on those areas that will have the greatest impact. A survey across multiple pharmaceutical companies of materials used to manufacture an API[2] revealed solvents to be the largest component, at 56% by mass. A similar analysis of medicinal chemistry scale reactions might be expected to give an even greater percentage, not only because chromatography is commonly used for purification, but also because the small scale of many medicinal chemistry reactions typically results in reactions being run in a less concentrated solution for ease of handling.

The criteria that are generally considered when assessing the environmental credentials of a solvent include:

- Carbon footprint
- Biodegradability
- Solubility in water (solvents that are immiscible with water are less likely to contaminate aqueous waste streams)
- Whether it gives rise to volatile organic compounds (VOCs) (solvents with higher boiling points are less likely to have this issue)
- Whether the solvent is easy to recycle (distillation is less energy intensive for lower boiling solvents)
- Whether the solvent is produced from a renewable feedstock

Additionally, safety criteria can be incorporated, including the solvent's:

- Stability
- Flammability (or flash point) and
- Toxicity (human and environmental)

Solvent guides can also be used to flag legislative issues, such as category 1A and 1B carcinogenic, mutogenic or reprotoxic (CMR) solvents,[3] or substances that deplete the ozone layer.[4]

There is no universal approach to solvent selection. A solvent that has the best sustainability credentials for one application (*e.g.* where recyclability or avoiding contamination of an aqueous waste stream is paramount) may not be the appropriate choice for another use (where avoiding VOCs or a requirement for compatibility with aqueous media may be a priority). Accordingly, in addition to providing at-a-glance guidance, solvent guides seek to explain the rationale behind the ranking of different solvents, providing the chemists with the information they need to make the right choice for a given task.

A number of solvent guides have been prepared to give guidance. Several pharmaceutical companies, including AstraZeneca, have produced in-house guides and some companies have published their guides. Pfizer[5] was the first to share their guide directed towards medicinal chemists, and in addition to providing a simplified solvent selection guide categorising solvents commonly used in medicinal chemistry as "preferred", "usable" and "undesirable", they also shared a solvent replacement table suggesting alternatives to the undesirable solvents (Figure 2.1).

The GlaxoSmithKline solvent selection guide[6] aimed to allow objective assessment of solvents and facilitate decisions as to whether existing or new "greener" solvents were really a more sustainable choice or whether they are just addressing a single sustainability issue. To allow this comparison, each solvent is scored from 1 (red) to 10 (green) to give a relative ranking for each solvent in the following categories:

- Waste: including recycling, incineration, VOC and biotreatment issues
- Environmental impacts: including the fate and effects of solvents on the environment
- Health: including acute and chronic effects on human health and the potential for exposure
- Flammability and explosion: issues affecting storage and handling of solvents
- Reactivity and stability: including factors affecting the stability of the solvent
- Life cycle: including the environmental life cycle impacts from producing a solvent
- Regulatory flag: alerting users of potential impacts from current or possible future environmental, health and safety related legislation
- Melting and boiling points: Red flags are attached to high (>120 °C) and low boiling point (<40 °C) solvents, with the aim of discouraging scientists from choosing solvents that have high energy requirements for separation by distillation, or solvents that may be highly volatile at ambient temperature and pressure. Melting points allow rapid identification of solvents that may be solid at room temperature.

The medicinal chemistry version of the GlaxoSmithKline solvent guide[6] was designed to act as an "at-a-glance" reference. Whilst not as comprehensive as the full 110 solvent guide,[6] it includes the 45 solvents judged to be most relevant to a medicinal chemistry setting. The front of the guide (Figure 2.2) collates solvents by category and allows the user to read horizontally across from red to green to search for more sustainable replacement solvents that are likely to be suitable. The reverse of the guide (Figure 2.3) contains further detail for each solvent listed, similar to in the full guide.

Sanofi's solvent selection guide provides safety, health and environmental scores, as well as the main industrial issues linked to the use of the solvent. Solvents are ranked into four categories: recommended, substitution

Preferred	Usable	Undesirable
Water	Cyclohexane	Pentane
Acetone	Heptane	Hexane(s)
Ethanol	Toluene	Di-isopropyl ether
2-Propanol	Methylcyclohexane	Diethyl ether
1-Propanol	Methyl *t*-butyl ether	Dichloromethane
Ethyl acetate	Isooctane	Dichloroethane
Isopropyl acetate	Acetonitrile	Chloroform
Methanol	2-MethylTHF	Dimethyl formamide
Methyl ethyl ketone	Tetrahydrofuran	*N*-Methylpyrrolidinone
1-Butanol	Xylenes	Pyridine
t-Butanol	Dimethyl sulfoxide	Dimethyl acetate
	Acetic acid	Dioxane
	Ethylene glycol	Dimethoxyethane
		Benzene
		Carbon tetrachloride

Undesirable solvents	Alternative
Pentane	Heptane
Hexane(s)	Heptane
Di-isopropyl ether or diethyl ether	2-MeTHF or *tert*-butyl methyl ether
Dioxane or dimethoxyethane	2-MeTHF or *tert*-butyl methyl ether
Chloroform, dichloroethane or carbon tetrachloride	Dichloromethane
Dimethyl formamide, dimethyl acetamide or *N*-methylpyrrolidinone	Acetonitrile
Pyridine	Et_3N (if pyridine used as base)
Dichloromethane (extractions)	EtOAc, MTBE, toluene, 2-MeTHF
Dichloromethane (chromatography)	EtOAc/heptane
Benzene	Toluene

Figure 2.1 The Pfizer solvent guide and solvent replacement table.
Reproduced from ref. 5 with permission of The Royal Society of Chemistry.

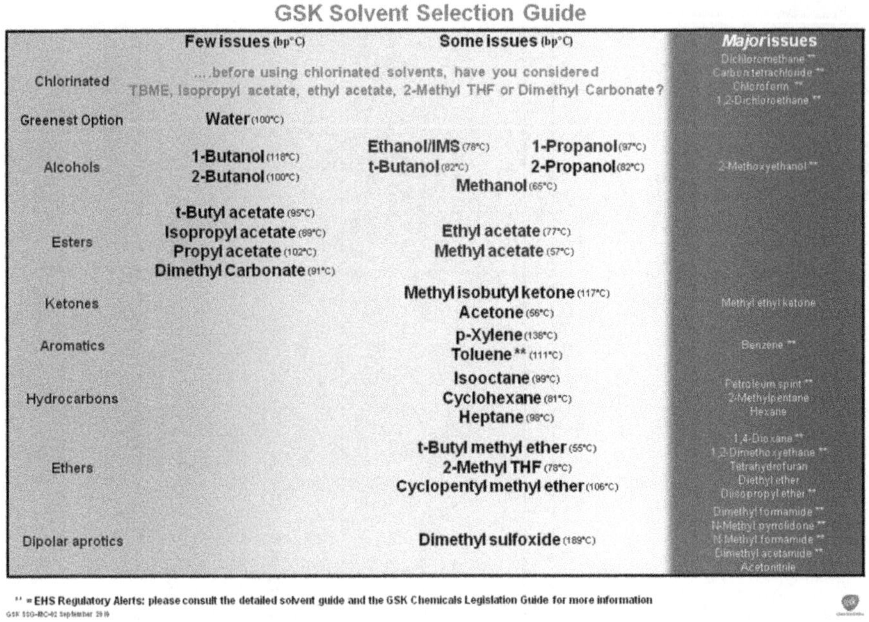

Figure 2.2 The front of the abbreviated medicinal chemistry version of the GSK solvent guide.
Reproduced from ref. 6 with permission of The Royal Society of Chemistry.

advisable, substitution requested and banned.[7] A table is proposed for each family of solvents, from which chemists can rapidly select the most sustainable candidate (Figure 2.4).

Any solvent guide is only as good as the data that has gone into its construction, and areas where information is currently unavailable should be highlighted. The guides should be treated as living documents and updated regularly as new toxicology and Environment Health and Safety (EHS) data become available.

Some work is ongoing in order to unify the solvent selection guides. The American Chemical Society Green Chemistry Institute Pharmaceutical Roundtable (ACS GCI PR) sought to provide one unified guide[8] that anyone could use. Their version covers 63 common solvents, and is based on the assessment of 5 categories: safety, health, environment (air, water, waste). CHEM21 is also actively working in this field. The data given in the publically available solvent selection guides were combined in order to allow a ranking comparison.[9] Of the 51 solvents considered, an acceptable alignment of the classification could be met for 67% of them (Figure 2.5). The others could not be unequivocally ranked, thus reflecting differences in the weighting of criteria between the institutions. The CHEM21 solvent subteam has since worked on the elaboration of a ranking methodology and the integration of bio-derived solvents.[10]

Classification	Solvent	CAS number	Melting point °C	Boiling point °C	Waste recycling, incineration, VOC, and biotreatment issues	Environmental Impact fate and effects on the environment	Health acute and chronic effects on human health and exposure potential	Flammability & Explosion storage and handling	Reactivity/ Stability factors affecting the stability of the solvent	Life Cycle Score Environmental impacts to produce the solvent	Legislation Flag alerts regulatory restrictions
Greenest	Water	7732-18-5	0	100	4	10	10	10	10	10	
Alcohols	1-Butanol	71-36-3	-89	118	5	7	5	8	9	5	
	2-Butanol	78-92-2	-115	100	4	6	8	7	9	6	
	Ethanol/IMS	64-17-5	-114	78	3	8	8	6	9	9	
	t-Butanol	75-65-0	25	82	3	9	6	6	10	8	
	Methanol	67-56-1	-98	65	4	9	5	5	10	9	
	2-Propanol	67-63-0	-88	82	3	9	8	6	8	4	
	1-Propanol	71-23-8	-127	97	4	7	5	7	10	7	
	2-Methoxyethanol	109-86-4	-85	124	3	8	2	7	6	7	
Ester	t-Butyl acetate	540-88-5	-78	95	6	9	8	6	10	8	
	Isopropyl acetate	108-21-4	-73	89	5	7	7	6	9	7	
	Propyl acetate	109-60-4	-92	102	5	7	8	6	10	4	
	Dimethyl carbonate	616-38-6	-1	91	4	8	7	6	10	8	
	Ethyl acetate	141-78-6	-84	77	4	8	8	4	8	6	
	Methyl acetate	79-20-9	-98	57	3	9	7	4	9	7	
Ketone	Methylisobutyl ketone	108-10-1	-84	117	6	6	6	7	8	2	
	Acetone	67-64-1	-95	56	3	9	8	4	9	7	
	Methylethyl ketone	78-93-3	-87	80	3	7	8	4	8	3	
Organic Acids	Propionic acid	79-09-4	-21	141	4	8	6	8	8	7	
	Acetic acid (glacial)	64-19-7	17	118	4	8	6	8	7	8	
Aromatics	p-Xylene	106-42-3	-13	138	7	2	6	5	10	7	
	Toluene	108-88-3	-95	111	6	3	4	4	10	7	
	Benzene	71-43-2	6	80	5	6	1	3	10	7	
Hydrocarbons	Isooctane	540-84-1	-107	99	6	4	8	3	10	7	
	Cyclohexane	110-82-7	7	81	5	5	7	2	10	7	
	Heptane	142-82-5	-91	98	6	3	8	3	10	7	
	2-Methylpentane	107-83-5	-153	60	5	4	7	2	10	7	
	Hexane	110-54-3	-95	69	5	3	4	2	10	7	
	Petroleum spirit	8032-32-4	-73	55	6	2	2	3	10	7	
Ethers	t-Butyl methyl ether	1634-04-4	-109	55	4	5	5	3	9	8	
	Cyclopentyl methyl ether	5614-37-9	-140	106	6	4	4	5	8	4	
	2-Methyl THF	96-47-9	-137	78	4	5	4	3	6	4	
	Diethyl ether	60-29-7	-116	35	4	4	5	2	4	6	
	Bis(2-methoxyethyl) ether	111-96-6	-68	162	4	5	2	8	4	6	
	1,4-Dioxane	123-91-1	12	102	3	4	4	4	5	6	
	Tetrahydrofuran	109-99-9	-108	65	3	5	6	3	4	4	
	1,2-Dimethoxyethane	110-71-4	-58	85	4	5	2	4	4	7	
	Diisopropyl ether	108-20-3	-86	68	4	3	8	1	1	9	
Dipolar aprotics	Dimethyl sulfoxide	67-68-5	19	189	5	5	7	9	2	6	
	Dimethyl formamide	68-12-2	-61	153	4	6	2	9	9	7	
	N-Methylformamide	123-39-7	-4	200	4	6	2	10	10	7	
	N-Methyl pyrrolidone	872-50-4	-24	202	5	6	3	9	8	4	
	Dimethyl acetamide	127-19-5	-20	165	5	6	2	10	8	2	
	Acetonitrile	75-05-8	-45	82	2	6	6	6	10	3	
Chlorinated	Carbon tetrachloride	56-23-5	-23	77	4	5	3	4	10	7	
	Dichloromethane	75-09-2	-95	40	3	6	4	6	9	7	
	Chloroform	67-66-3	-64	61	3	6	3	6	9	6	
	1,2-Dichloroethane	107-06-2	-36	84	4	4	2	6	10	7	

Legislation Flag

- Substitution recommended - There are no current restrictions but future regulatory restrictions may apply
- Substitution recommended - existing regulatory restrictions apply
- Must be substituted - A regulatory ban applies

Figure 2.3 The back of the abbreviated medicinal chemistry version of the GSK solvent guide. Reproduced from ref. 6 with permission of The Royal Society of Chemistry.

Solvents guide

APROTIC POLAR SOLVENTS: OVERVIEW

SANOFI

Name	Overall evaluation	ICH limit (ppm)	Occ. health	Safety	Environment	Other concern
Acetonitrile	Recommended	410	OEBV3 Sk	SHB3	EHB1	Availability
Propionitrile	Substitution requested	Not listed	OEBV4	SHB3	Not available	Decomposition gives HCN
Dimethyl-formamide	Substitution requested	880	OEBV4 G2Sk	SHB2	EHB1	CMR (reprotoxic)
Dimethyl-acetamide	Substitution requested	1090	OEBV4 G2Sk	SHB2	EHB1	CMR (reprotoxic)
N-Methyl-pyrrolidone	Substitution requested	530	OEBV3 G2Sk	SHB3	EHB1	CMR (reprotoxic); high BP
Dimethyl-ethylene-urea	Substitution advisable	Not listed	OEBV2 G1	SHB2	EHB2	Hygroscopic, high BP
Dimethyl-propylene-urea	Substitution advisable	Not listed	OEBV3 G1	SHB3	EHB2	Hygroscopic, high BP, cost
Dimethyl-sulfoxide	Substitution advisable	5000	OEBV2 Sk	SHB4	EHB2	Many incompatibilities, odor
Sulfolane	Substitution advisable	160	OEBV2	SHB1	EHB2	High BP and MP
Nitromethane	Banned	50	OEBV3	SHB5	EHB3	High energy of decomposition

Figure 2.4 Table of aproptic polar solvents in Sanofi's guide.
Reprinted with permission from ref. 7. Copyright 2013 American Chemical Society.

Recommended	Water, EtOH, i-PrOH, *n*-BuOH, EtOAc, i-PrOAc, *n*-BuOAc, anisole, sulfolane.
Recommended or problematic?	MeOH, *t*-BuOH, benzyl alcohol, ethylene glycol, acetone, MEK, MIBK, cyclohexanone, MeOAc, AcOH, Ac_2O.
Problematic	Me-THF, heptane, Me-cyclohexane, toluene, xylenes, chlorobenzene, acetonitrile, DMPU, DMSO.
Problematic or hazardous?	MTBE, THF, cyclohexane, DCM, formicacid, pyridine.
Hazardous	Diisopropyl ether, 1,4-dioxane, DME, pentane, hexane, DMF, DMAc, NMP, methoxy-ethanol, TEA.
Highly hazardous	Diethyl ether, benzene, chloroform, CCl_4, DCE, nitromethane.

Figure 2.5 Overall ranking of classical solvents.
Reproduced from ref. 9 with permission of The Royal Society of Chemistry.

Additional tools can help to break down perceived barriers to the usage of some of the more unfamiliar solvents. Building on the frequent use made of lists of peaks arising from trace residual solvents in common deuterated NMR solvents,[11] NMR guides[12] emphasising the peaks of some of the less familiar and potentially greener solvents in common deuterated NMR solvents (chloroform, DMSO, methanol, water, acetone and acetonitrile), can help act as a reminder of different options and help remove some of the barriers to their uptake.

Whilst the above guides provide general information on a large number of solvents, for some purposes more detailed information is required.

2.2.1 Solvent Guidance for Chromatography

Chromatography is one of the largest contributors to solvent waste in medicinal chemistry, and a number of alternatives and modifications to flash chromatography have been considered.[13]

The aforementioned guides do not allow for ready comparison of the polarity of different solvents. Equielutropic series have a long history,[14] and have been used in medicinal chemistry laboratories for decades. Recent versions have focussed on highlighting more sustainable alternatives to dichloromethane-based eluants.

Allan Watson[15] at the University of Strathclyde evaluated a number of alternative solvents as potential replacements for dichloromethane as the major component of the eluant for chromatographic purification. The performance was benchmarked over a range of both polar fragments and more complex molecules. Box plot analysis of the retardation factor (R_fs) observed allowed for broad comparison between solvent systems.

Amgen[16] used a set of "drug-like" compounds with a range of functional groups and physicochemical parameters to compare the relative eluting strengths of greener solvent systems. Their user-friendly guide allows ready comparison of solvent systems of similar eluting ability across neutral acidic and basic compound categories. In particular, they found a gradient of a 3 : 1 mixture of ethyl acetate : ethanol in heptanes to cover the range of polarity historically covered by a gradient of methanol in dichloromethane. Premixed 3 : 1 ethyl acetate : ethanol is now being supplied by several solvent retailers, reducing the barrier to adopting this approach (Figure 2.6).

2.2.2 Solvent Guidance for Principal Component Analysis (PCA) Viewers

Design of Experiment (DoE) approaches (see Chapter 10) enable the screening of different reaction conditions. Solvent maps are commonly used as a tool in experimental design, for example, where one solvent out of each 8 "octants" might be selected in order to give coverage of a broad range of physicochemical space. Inclusion of scoring or colour coding for the solvents included in these solvent maps can enable the selection of more sustainable solvents right from the start of a screening process.

2.2.3 Tools That can Minimise Reliance on Non-preferred Solvents

2.2.3.1 Hydrophilic Frits

Whilst not as great as the volume of solvent that can be used in purification, the solvent used for work-up or extraction after a reaction is complete can still be many times the volume of the reaction solvent. For individual reactions on a medicinal chemistry scale, separating funnels are commonly used and solvents such as ethyl acetate or 2-methyl tetrahydrofuran can be recommended for extractions. (Indeed, the immiscibility of 2-methyl tetrahydrofuran with water represents a significant advantage over tetrahydrofuran for some reactions for this very reason.) However, array work has increasingly involved the use of hydrophobic frits allowing the rapid separation of denser than water chlorinated solvents from an aqueous layer. Similarly hydrophilic frits allow an aqueous layer to be run off to waste, permitting the less dense organic solvent to be tipped into another receptacle for concentration.

2.2.3.2 Ion Exchange Chromatography

Ion exchange chromatography can allow for quicker, less sorbent- and solvent-intensive purification than flash chromatography in instances where the ionisation state of the product is differentiated from all other reaction components. It can also partially clean mixtures before other purification

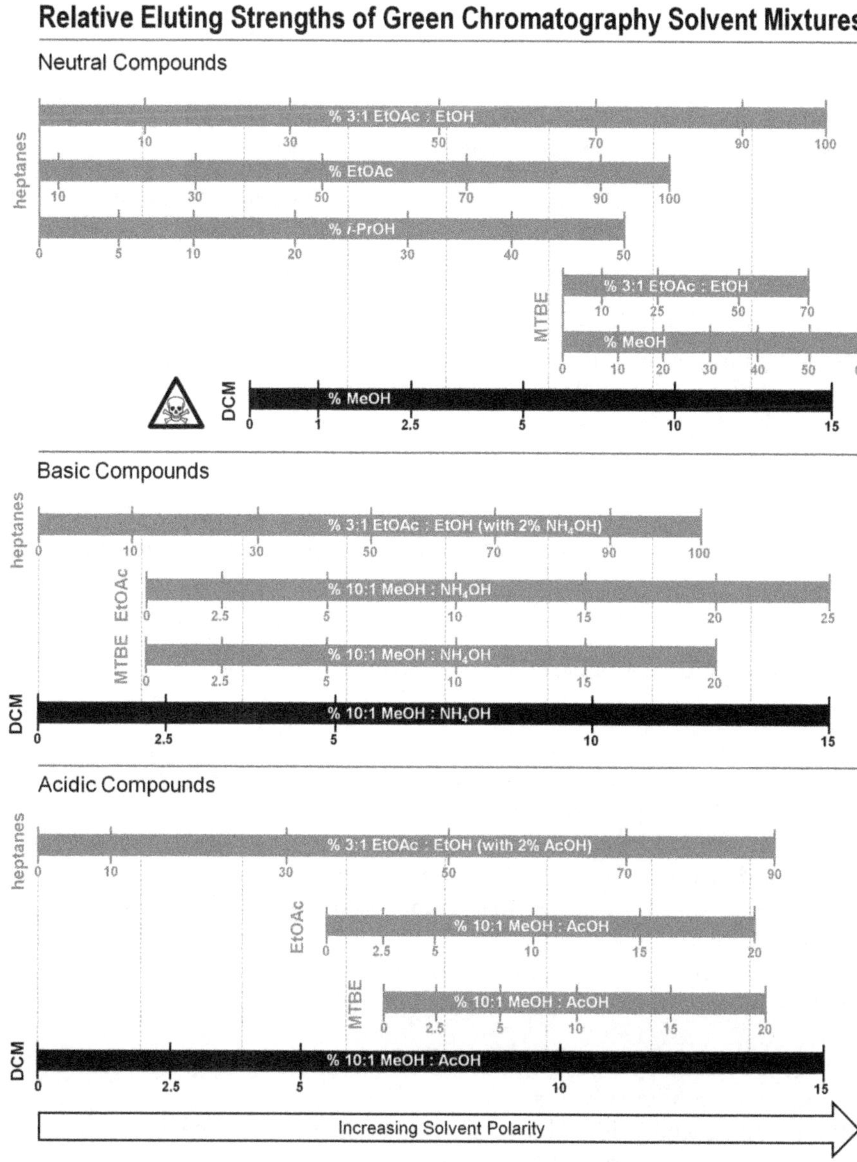

Figure 2.6 The Amgen green chromatography solvent selection guide. Reproduced from ref. 16 with permission of The Royal Society of Chemistry.

techniques are applied in complex cases, desalt compounds before reaction or remove acid modifiers after reverse phase HPLC (high-performance liquid chromatography).

Ion exchange chromatography relies on silica, modified by bonds to a variety of functional groups. It can be useful where a rapid workup is

Table 2.1 Ion exchange column guidance.

Mixture **Target** + *Impurity*	Cartridge type	Elution steps
Neutral + *Acidic*	Aminopropyl (NH2)	**Product elutes in MeOH wash**
Acidic + *Neutral*	Aminopropyl (NH2)	(i) *MeOH wash* (ii) **Product elutes in MeOH/HCl**
Neutral + *Basic*	Sulfonic acid (SCX-2)	**Product elutes in MeOH wash**
Basic + *Neutral*	Sulfonic acid (SCX-2)	(i) *MeOH wash* (ii) **Product elutes in MeOH/NH_3**

essential, especially where decomposition is suspected for slower purification techniques, as crude reaction mixtures can be applied directly onto a pre-conditioned column.

The advantages offered by this approach must be balanced against the costs, both financial and environmental, of using a highly processed resource, and it is only applicable where ionisable and non-ionisable components require separation.

Simple guides (Table 2.1) can advise which approach can be used in each instance.

2.3 Reagent Selection Guides

Reagent selection guides can help in a number of ways. They can identify:

- Alternatives to reagents whose use may be becoming increasingly restricted by legislation
- Alternatives to reagents that might be constrained by company targets around reducing the use of materials of concern
- Reagents that have lesser EHS implications (by virtue of having more benign risk phrases)
- Reagents/procedures that are more atom efficient
- Reactions with fewer operational complexities

Reagent selection guides can also direct chemists to reagents that can be used in more benign solvents.

Pfizer introduced guides[5] based on a Venn diagram approach that allowed at-a-glance comparison of utility, scalability and "greenness", helping to bridge the gap between the needs of the medicinal chemist (where the capacity for a reagent to handle a range of substrates makes for streamlined research) and the needs of the development chemist where the ability to be safely, reliably and cost effectively handled on a large scale is paramount (Figure 2.7).

Recently, GlaxoSmithKline published reagent guides covering 15 of the most commonly used transformations in medicinal chemistry.[17] Each guide covers between 13 and 27 reagents colour-coded using the traffic light system with an overall greenness score (Figure 2.8). Further detail on the scores

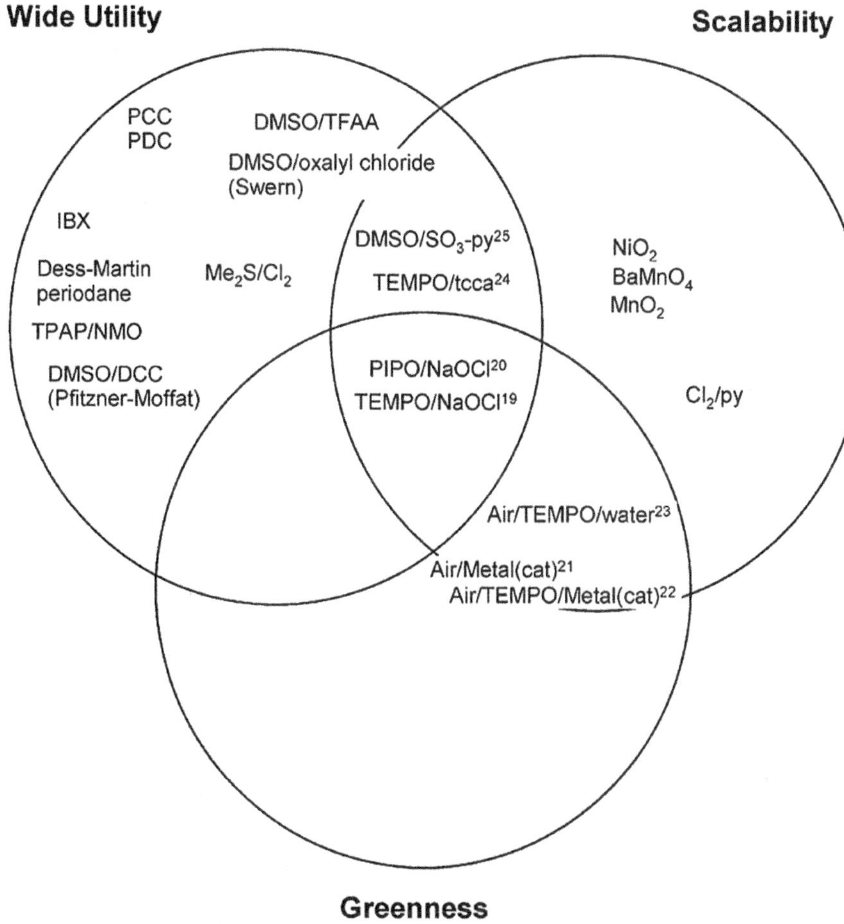

Figure 2.7 Sample Pfizer reagent guides for oxidation of primary alcohol to aldehyde.
Reproduced from ref. 5 with permission of The Royal Society of Chemistry.

for different categories (health, safety and environment scores based on risk phrases, chemistry scores based on work-up, co-reagent, atom efficiency, stoichiometry and other issues), which were combined to give that greenness score, are available on the reverse of the guides (Figure 2.9), alongside additional comments or advice.

For both the above guides, further details are available within each company, with links to additional information and sample protocols.

The ACS GCI PR (American Chemical Society Green Chemistry Institute Pharmaceutical Roundtable) is preparing a combined reagent guide, which will feature contributions from both the Pfizer and GSK guides mentioned and multiple other companies, with release scheduled in the near future. This guide will include references aligned to the greenest conditions,

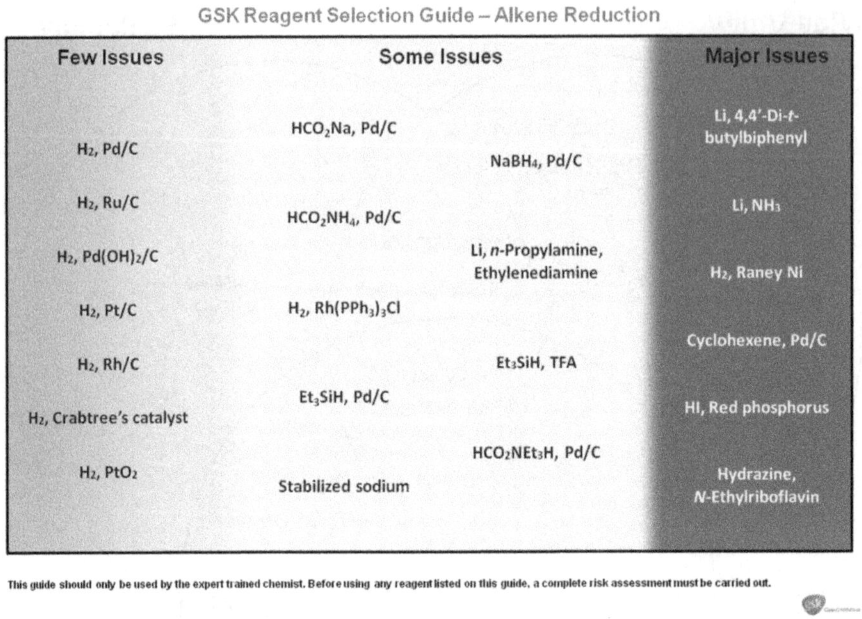

Figure 2.8 Sample GSK reagent guide for alkene reduction. Reproduced from ref. 17 with permission of The Royal Society of Chemistry.

examples used in scaled-up processes, discussion of environmental considerations, cost, atom efficiency, availability, awareness of sustainable feedstocks, reaction mechanisms, general hints and tips and a list of prominent references for each general transformation.

2.4 Reagent and Solvent Selection—Substrate Specificity

Allan Watson explored the substrate and solvent specificity of a range of common reagents for amide bond formation[18] and reductive amination[19] with a particular view to exploring alternatives to dichloromethane and dimethylformamide. Charts highlight which substrates, solvent and reagents are compatible.

In-house databases can go a long way towards providing relevant precedents for reactions in preferred solvents or with preferred reagents. However, to properly embed sustainability in medicinal chemistry, it would be helpful to be able to filter external databases by the solvents and reagents used. Reaxys has already shown that this is possible.

Electronic laboratory notebooks (eLNB or ELN) have enormous potential. Not only can they be mined to explore precedents when searching for more sustainable reaction conditions, but they can be set up in such a way as to encourage the consideration of sustainability right from the planning

Material	Full name	CAS / MFCD number	EHS	Clean chemistry	Greenness	Comment
H_2, Pd/C	Hydrogen, Palladium on Carbon	1333-74-0/7440-05-3	7	7	7.2	Flammable. Relative cost cheap. Catalyst can be recycled.
H_2, Ru/C	Hydrogen, Ruthenium on Carbon	1333-74-0/7440-18-8	7	7	7.2	Flammable. Relative cost cheap. Catalyst can be recycled.
H_2, $Pd(OH)_2$/C	Hydrogen, Palladium hydroxide	1333-74-0 12135-22-7	7	7	7.2	Flammable. Slightly more expensive than Pd/C. Catalyst can be recycled.
H_2, Pt/C	Hydrogen, Platinum on Carbon	1333-74-0 7440-06-4	7	7	7.2	Flammable. Relative cost expensive. Catalyst can be recycled.
H_2, Rh/C	Hydrogen, Rhodium on Carbon	1333-74-0/7440-16-6	7	7	7.1	Flammable. Relative cost extremely expensive.
H_2, Crabtree's catalyst	Hydrogen, (1,5-Cyclooctadiene)(pyridine)(tricyclohexylphosphine)-iridium(I) hexafluorophosphate	1333-74-0 64536-78-3	7	6	6.8	Flammable. Relative cost of iridium very expensive. Catalyst can be recycled.
H_2, PtO_2	Hydrogen, Platinum oxide	1333-74-0 1314-15-4	5	7	6.1	Flammable. Relative cost expensive. Catalyst can be recycled.
HCO_2Na, Pd/C	Sodium Formate, Palladium on Carbon	141-53-7 7440-05-3	7	5	5.9	Hydrogen transfer. Reaction rates, off-gassing rates may be a concern on some systems.
Ammonium formate, Pd/C	Ammonium Formate, Palladium on Carbon	540-69-2 7440-05-3	7	5	5.8	Hydrogen transfer. Reaction rates, off-gassing rates may be a concern on some systems.
H_2, $Rh(PPh_3)_3Cl$	Hydrogen, Chlorotris(triphenylphosphine)rhodium(I)	1333-74-0/14694-95-2	5	6	5.8	Flammable. Relative cost of rhodium extremely expensive.
Et_3SiH, Pd/C	Triethylsilane, Palladium on Carbon	617-86-7/7440-05-3	7	5	5.5	Hydrogen transfer.
Stabilized sodium	Sodium on Silica gel	MFCD01941718	5	5	5.3	Used in Birch reduction, no need for ammonia.
$NaBH_4$, Pd/C	Sodium borohydride, Palladium on Carbon, Acetic acid	16940-66-2 7440-05-3/64-19-7	5	5	5.3	Acetic acid required.
Li, *n*-Propylamine, $NH_2CH_2CH_2NH_2$	Lithium, *n*-Propylamine, Ethylenediamine	7439-93-2/107-10-8 107-15-3	5	5	5.0	Flammable.
Et_3SiH, TFA	Triethylsilane, Trifluoroacetic acid	617-86-7 76-05-1	5	5	4.9	Higher stoichiometry of Et_3SiH may be required for less electron rich alkenes.
HCO_2NEt_3H, Pd/C	Triethylammonium formate, Palladium on Carbon	585-29-5 7440-05-3	5	4	4.9	Hydrogen transfer. Reaction rates, off-gassing rates may be a concern on some systems.
Li, 4,4'-Di-*t*-butylbiphenyl	Lithium, 4,4'-Di-*tert*-butylbiphenyl	7439-93-2 1625-91-8	5	2	3.5	Causes burns. Reacts violently with water.
Li, NH_3	Lithium, Ammonia	7439-93-2 7664-41-7	4	3	3.3	Causes burns. Very toxic to aquatic organisms. Danger of serious irreversible effects by exposure.
H_2, Raney Ni	Hydrogen, Raney nickel	1333-74-0 7440-02-0	1	6	3.0	Highly flammable. Potential carcinogen and sensitizer.
Cyclohexene, Pd/C	Cyclohexene, Palladium on Carbon	110-83-8 7440-05-3	1	6	3.0	Benzene (known carcinogen) forms as by-product.
HI, Red phosphorus	Hydroiodic acid, Red phosphorus	10034-85-2 7723-14-0	2	3	2.7	Highly flammable. Red phosphorus is a controlled substance.
Hydrazine, *N*-Ethylriboflavin	Hydrazine, *N*-Ethylriboflavin	302-01-2 N/A	1	4	2.4	Hydrazine is a material of concern. May cause cancer.

Figure 2.9 Sample GSK reagent guide for alkene reduction.
Reproduced from ref. 17 with permission of The Royal Society of Chemistry.

of a reaction.[20] ELNBs can allow for the ready calculation and sharing of green metrics, such as *E*-factor[21] or Process Mass Intensity (PMI),[22] allowing improvements in routes to be readily tracked as projects progress. It is also conceivable that the solvent and reagent guides discussed above may also be integrated into eLNBs and templates for common reactions helping to influence decision making. In addition, the move to paperless working (which eLNBs represent just a small part of) is of itself to be commended for its improved sustainability credentials.

2.5 Future Opportunities

Green teams within various organisations and non-competitive consortia of pharmaceutical companies are working to both troubleshoot current environmental sustainability problem areas, and spot and promote best practices and future ways of working as they arise.[23,24]

2.6 Summary

- Detailed analysis or life cycle assessment of individual reactions or routes is not appropriate for medicinal chemistry, but the general principles can still inform medicinal chemistry choices.
- Solvent usage is the largest contributing factor to the environmental footprint of medicinal chemistry, and general solvent guidance is available.
- More specialised solvent guidance tools are available for chromatography, purification or reaction screening purposes.
- Reagent guides can give generalised advice on common reaction classes.
- More detailed advice on the scope and utility of specific reagents is slowly improving and can be accessed through databases and publications exploring the strengths and limitations of reagents.

References

1. J. G. Lombardino and J. A. Lowe, *Nat. Rev. Drug Discovery*, 2004, **3**, 853.
2. D. J. C. Constable, C. Jimenez-Gonzalez and R. K. Henderson, *Org. Process Res. Dev.*, 2007, **11**, 133.
3. According to European council directive 1999/38/EC, any solvent with the H350 (carcinogenic) or H360 (reprotoxic) hazard phrase in the Globally Harmonized System (GHS) of Classification and Labelling of Chemicals must be substituted, or have their use justified.
4. According to European parliament regulation 1005/2009, the use of any solvent with the H420 (harms public health and the environment by destroying ozone in the upper atmosphere) hazard phrase in the Globally Harmonized System (GHS) of Classification and Labelling of Chemicals is severely restricted.

5. K. Alfonsi, J. Colberg, P. J. Dunn, T. Fevig, S. Jennings, T. A. Johnson, H. P. Kleine, C. Knight, M. A. Nagy, D. A. Perry and M. Stefaniak, *Green Chem.*, 2008, **10**, 31.
6. R. K. Henderson, C. Jiménez-González, D. J. C. Constable, S. R. Alston, G. G. A. Inglis, G. Fisher, J. Sherwood, S. P. Binks and A. D. Curzons, *Green Chem.*, 2011, **13**, 854.
7. D. Prat, O. Pardigon, H. W. Flemming, S. Letestu, V. Ducandas, P. Isnard, E. Guntrum, T. Senac, S. Ruisseau, P. Cruciani and P. Hosek, *Org. Process Res. Dev.*, 2013, **17**, 1517.
8. http://surveys.acs.org/se.ashx?s=04BD76CC0E5496A7.
9. D. Prat, J. Hayler and A. Wells, *Green Chem.*, 2014, **16**, 4546.
10. D. Prat, A. Wells, J. Hayler, H. Sneddon, C. R. McElroy, S. Abou-Shehada and P. J. Dunn, *Green Chem.*, 2016, **18**, 288; D. Prat, A. Wells, J. Hayler, H. Sneddon, C. R. McElroy, S. Abou-Shehada and P. J. Dunn, *Green Chem.*, 2015, **17**, 4848.
11. *e.g.* H. E. Gottlieb, V. Kotlyar and A. Nudelman, *J. Org. Chem.*, 1997, **62**, 7512.
12. H. E. Gottlieb, G. Graczyk-Millbrandt, G. G. A. Inglis, A. Nudelman, D. Perez, Y. Qian, L. E. Shuster, H. F. Sneddon, R. J. Upton, Manuscript in preparation.
13. E. A. Peterson, B. Dillon, I. Raheem, P. Richardson, D. Richter, R. Schmidt and H. F. Sneddon, *Green Chem.*, 2014, **16**, 4060.
14. R. Neher, in *Thin Layer Chromatography*, ed. G. B. Marini-Bettolo, Elsevier, Amsterdam, 1964, p. 77.
15. D. S. MacMillan, J. Murray, H. F. Sneddon, C. Jamieson and A. J. B. Watson, *Green Chem.*, 2012, **14**, 3016.
16. J. P. Taygerly, L. M. Miller, A. Yee and E. A. Peterson, *Green Chem.*, 2012, **14**, 3020.
17. J. P. Adams, C. M. Alder, I. Andrews, A. M. Bullion, M. Campbell-Crawford, M. G. Darcy, J. D. Hayler, R. K. Henderson, C. A. Oare, I. Pendrak, A. M. Redman, L. E. Shuster, H. F. Sneddon and M. D. Walker, *Green Chem.*, 2013, **15**, 1542.
18. D. S. MacMillan, J. Murray, H. F. Sneddon, C. Jamieson and A. J. B. Watson, *Green Chem.*, 2013, **15**, 596.
19. F. I. McGonagle, D. S. MacMillan, J. Murray, H. F. Sneddon, C. Jamieson and A. J. B. Watson, *Green Chem.*, 2013, **15**, 1159.
20. M. E. Kopach and E. A. Reiff, *Future Med. Chem.*, 2012, **4**(11), 1395.
21. R. A. Sheldon, *Green Chem.*, 2007, **9**, 1273.
22. C. Jimenez-Gonzalez, C. S. Ponder, Q. B. Broxterman and J. B. Manley, *Org. Process Res. Dev.*, 2011, **15**(4), 912.
23. M. C. Bryan, L. Hamann, G. Hughes, M. Kopach, E. Peterson, M. Pourashraf, I. Raheem, P. Richardson, D. Richter and H. F. Sneddon, *J. Med. Chem.*, 2013, **56**, 6007.
24. D. K. Leahy, J. L. Tucker, I. Mergelsberg, P. J. Dunn, M. E. Kopach and V. C. Purohit, *Org. Process Res. Dev.*, 2013, **17**(9), 1099.

CHAPTER 3

Renewable Solvent Selection in Medicinal Chemistry

JAMES H. CLARK,* ANDREW J. HUNT, LAURIANNE MOITY AND JAMES SHERWOOD

Green Chemistry Centre of Excellence, Department of Chemistry, University of York, York YO10 5DD, UK
*Email: james.clark@york.ac.uk

3.1 Sources of Bio-based Solvents

Considerable pressure is mounting on chemical manufacturers to consider replacing processes that are dependent on non-renewable feedstocks. Alternative practices based on abundant plant-derived starting materials or the waste streams from other industries are becoming more attractive. For the greater part of the chemical industry's history, solvents have been made from fossil materials. A recent drive to establish fermentation routes to platform molecules,[1] and then onto solvents, is seeing a return to popularity for these traditional processes that predate the dominance of crude oil and natural gas as chemical feedstocks. For example, fermentation of sugars to give acetone, 1-butanol, and ethanol as joint products was popular in the past when availability and price were an issue. Recently, it has again become a commercially attractive process in order to satisfy the growing interest in 1-butanol as a fuel, and of course because of continued geopolitical uncertainties surrounding petroleum feedstock security.[2] The reason for focusing on the sustainability and environmental impact of solvents (and not the more obvious target of reactants) is the volume in which solvents are

RSC Green Chemistry No. 46
Green and Sustainable Medicinal Chemistry: Methods, Tools and Strategies for the 21st Century Pharmaceutical Industry
Edited by Louise Summerton, Helen F. Sneddon, Leonie C. Jones and James H. Clark

Published by the Royal Society of Chemistry, www.rsc.org

used. This dwarfs the mass contribution of other chemicals at all levels of chemistry in the pharmaceutical industry (Figure 3.1).[3]

Fermentation is a viable means of producing solvents, with various alcohols and acids produced commercially by this means. From bio-ethanol, the following familiar solvents have been made: acetic acid,[4] acetonitrile,[5] ethyl acetate,[6] ethyl lactate,[7] and triethylamine (Scheme 3.1).[8] The products of fermentation can also serve as platform molecules to deliver an even greater diversity of bio-based chemicals. Bio-ethanol is being dehydrated to give a renewable source of ethylene.[9] It is not hard to hypothesise that routes to further solvents that currently depend on petroleum derived ethylene, such

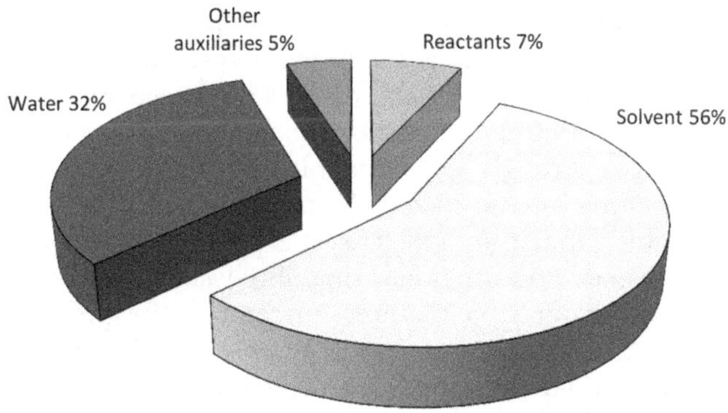

Figure 3.1 The relative contributions of materials typical in the production of an active pharmaceutical ingredient.[3]

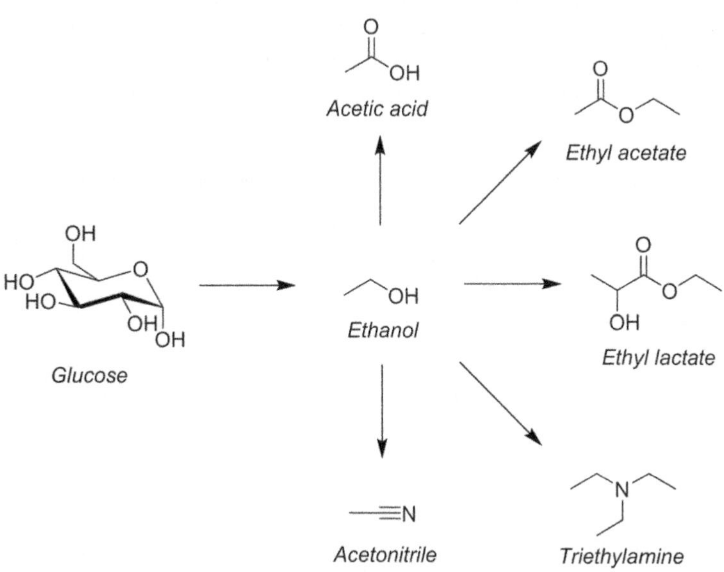

Scheme 3.1 Synthesis of solvents from bio-ethanol.

as diethyl ether, might have their current feedstock replaced with its plant based analogue in time.

A policy of replacing the petrochemical feedstocks for conventional solvents has the benefit of hugely simplifying solvent substitution. The problem with replacing petroleum derived solvents with the same bio-based solvent is that any safety or toxicity issues are not resolved. Environmental issues occurring at the end of use will also persist. With the REACH European regulation starting to influence solvent selection,[10] manufacturers will be forced to investigate alternative solvents. At least bio-based solvents are compatible with the development of environmentally sustainable processes.

Regarding the origin of bio-based solvents, it is important that food supply chain wastes should be the source of chemical products where at all possible.[11] This is because two substantial issues detract from the advantages of solvent substitution in favour of sugar derived bio-based solvents, especially those made by fermentation. This feedstock competes with our food supply, therefore creating a strongly objectionable conflict. Extending this argument, non-food crops for use in the chemical feedstock or biofuel sectors also require arable land, thus still creating pressure on food production (as well as biodiversity and other sustainability issues).[12] Secondly, the salt formation or distillation step required to remove the product from the fermentation broth is wasteful or energy intensive and results in aqueous waste that must be treated appropriately. The development of fermentation processes that use cellulose as a feedstock, and non-fermentation routes to bio-based solvents, need to be further developed to resolve these shortfalls.

An alternative to bio-based solvent production by fermentation is the anaerobic digestion of organic (food) wastes.[13] Solvents that are currently made from natural gas could quite easily be synthesised from the methane produced by anaerobic digestion (Scheme 3.2). A co-product of anaerobic digestion is carbon dioxide, which can also be used as a solvent.

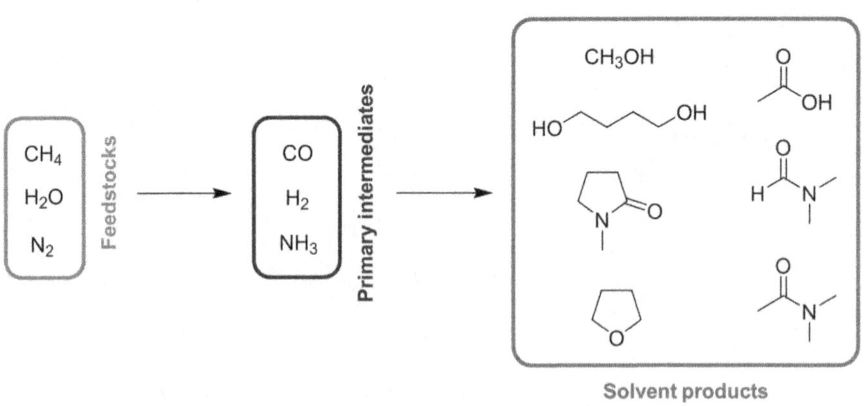

Scheme 3.2 Examples of common solvents made from methane.

This additional use of what is usually considered a waste product diminishes its environmental impact.

Carbon dioxide (utilised as a supercritical fluid) is one of several so-called neoteric solvent systems that have accelerated interest in solvents with the onset of green chemistry as a scientific discipline.[14] A supercritical fluid is defined as a substance that is above its critical temperature and critical pressure. This critical point occurs at 73.8 bar and 304.2 K for carbon dioxide.[15] The physical properties of a supercritical fluid are generally described as being between those of gases and liquids, although these properties can vary significantly as the temperature and pressure are changed within the system. The viscosities of supercritical fluids are typically an order of magnitude lower than those of liquids, whereas diffusivity is at least one order of magnitude higher. The density of a supercritical fluid is a key property in controlling the solubility of substrates; therefore, the solvency power of supercritical carbon dioxide ($scCO_2$) can be tuned by changing the temperature and pressure of the system. This feature of $scCO_2$ is useful for chromatography.[16]

Supercritical carbon dioxide has been utilised in a wide variety of applications in addition to chromatography, including materials preparation, natural product extraction, and in reaction chemistry.[14] After the process, be it reaction, extraction or separation, the carbon dioxide solvent can be completely removed simply by depressurising the supercritical fluid apparatus. This alleviates the issue of solvent contamination, which is a significant regulatory concern when it comes to the production of food grade chemicals and pharmaceutical products.[17] The apparatus required to produce and utilise $scCO_2$ are now successfully used in both academia and industry in chromatographic applications,[16] and in the extraction of natural products from biomass.[14,18] $scCO_2$ is beginning to attract interest in medicinal chemistry laboratories and is a promising technique deserving of future development.

Another example of a successful renewable solvent that is not a fermentation product is 2-methyltetrahydrofuran (2-MeTHF). It is commercially synthesised by the acid mediated dehydration of hemicellulose into furfural, which is followed by a hydrogenation.[19] Although similar to tetrahydrofuran (THF), the most obvious candidate for substitution by 2-MeTHF, it is a distinct molecule made by a process that has no equivalent downstream of the oil refinery where THF originates. Peroxide formation issues are unresolved, but 2-MeTHF does have advantages over THF. 2-Methyltetrahydrofuran is less water soluble than THF, and the achievable concentration of organometallic substrates is greater in 2-MeTHF than it is in THF.[19,20]

Some bio-based solvents require no synthesis at all, either because they exist in nature, or they are formed as a waste or by-product of another chemical process. Terpenes isolated by steam distillation or mechanical pressing of food waste can make useful solvents. Limonene from waste orange peel has been shown to be a high performance solvent for esterifications and amidations.[21] Oxidised to *p*-cymene, limonene provides the basis of a replacement for toluene and other popular aromatic solvents.[22,23] Glycerol, which might have been otherwise overlooked as a solvent because of its viscosity and high boiling point, has become a viable solvent option with the growth of bio-diesel production globally.[24] Glycerol is a value adding by-product of this process in the same way that carbon dioxide can be viewed as an additional product of anaerobic digestion or fermentation. The scope of glycerol (and its ester, ether, and carbonate derivatives) as a solvent is becoming quite broad, including Michael addition reactions, palladium catalysed cross-coupling, and heterocycle syntheses.[25]

3.2 Solvent Selection

Motivated to obtain high yields in short reaction times to give a high purity product, for the medicinal chemist solvent selection is often intuitive. Through a search of the scientific literature, and insight built through years of practice, a suitable solvent is rarely hard to come by for whatever the chosen reaction may be. However, the ideology that underpins the solvent selection that occurs in medicinal chemistry has its foundations in proven solvents from a past era. These conventional solvents present a low risk in terms of reaction failure, but solvents like chloroform and toluene present high risks to health and the environment. This conflicts with the needs of the process development chemists that take on the work of the medicinal chemists, for they cannot simply repeat their reactions on a larger scale. Process chemists must prioritise health and safety considerations, and give greater consideration to regulatory issues. Then in turn the pilot plant engineers must create a viable manufacturing procedure. Contrary to medicinal chemistry practice, now the end-of-life issues of solvent disposal and possible air pollution, water contamination, and environmental persistency are of a high priority. The reasons for choosing one solvent over another can seem far removed from the original intentions of reaction productivity.

This process of solvent substitution could be streamlined by establishing policies of safe (and ideally renewable) solvent use at the earliest stages of drug development to aid scale up. At some point a solvent must be selected and then used indefinitely, for once a process is registered it cannot be easily changed.[26] A solvent substitution near the end of a synthetic route would have huge implications for the final product, with different traces of impurities (and of course residual solvent) considered as modifying the substance from that which was originally registered.[27] As such, it appears that medicinal chemists have good reason to embrace a greater role in shaping future solvent preferences.

Finally, it is important to remember that the catalogue of solvents is not immutable. Different solvents naturally fall in and out of favour, depending on environmental regulations and the types of chemistry being performed at a given time. Data from GSK suggests that toluene, dichloromethane (DCM)

Table 3.1 The popularity of solvents at pilot plant scale within GSK.[28]

Solvent	Rank (2005)	Rank (1990–2000)
Isopropanol	1 (+4)	5
Ethyl acetate	2 (+2)	4
Methanol	3 (+3)	6
Ethanol	4 (+4)	8
n-Heptane	5 (+7)	12
THF	6 (−4)	2
Toluene	7 (−6)	1
DCM	8 (−6)	2
Acetic acid	9 (+2)	11
Acetonitrile	10 (+4)	14

and THF have dropped in popularity at pilot plant scale in recent times, which should be noted by medicinal chemists in their own solvent selection (Table 3.1).[28]

3.3 Solvent Polarity

3.3.1 Solubility Parameters

For all solvent candidates that are compatible with a given reaction, it is their polarity that will determine the performance of that reaction. Sadly, the term is vague in nature, difficult to define, and as a consequence poorly understood. The single parameter that many imagine as the basis of solvent polarity is dielectric constant (which in fact is an out-dated term for relative permittivity), or the slightly more sophisticated Hildebrand solubility parameter. The latter is defined as the square root of a solvent's cohesive energy density, the macroscopic manifestation of all the solvent–solvent intermolecular forces present.[29] The Hildebrand solubility parameter can be deconstructed into three separate functions to describe solvent dispersion forces (δ_D), dipolarity (δ_P), and hydrogen bonding (δ_H) as individual expressions, known as the Hansen solubility parameters (HSPs) in honour of their originator (Table 3.2).[30] Essentially a multi-parameter version of the old 'like-dissolves-like' adage, solvents that have similar HSPs to a solute would be expected to dissolve that compound. By attributing polarity across three parameters and not just one, the specifics of solubility can be better understood, and suitable green solvent substitutes more easily found.

However, some limitations of these parameters should be pointed out: firstly, HSPs have been experimentally measured for only 88 solvents.[30] Thus, most of the HSPs of solvents have been calculated computationally. Various calculation methods exist (group contribution method, neural networks, *etc.*) but still sometimes they can lead to inconsistent values. Secondly, the interpretation of HSPs only gives a qualitative indication of solubility (soluble or not soluble) instead of achievable substrate concentrations.

Table 3.2 Examples of Hansen solubility parameters for common solvents.[30]

Solvent	δ_D/MPa$^{1/2}$	δ_P/MPa$^{1/2}$	δ_H/MPa$^{1/2}$
Acetic acid	14.5	8.0	13.5
Acetone	15.5	10.4	7.0
Acetonitrile	15.3	18.0	6.1
Benzene	18.4	0.0	2.0
1-Butanol	16.0	5.7	15.8
Chloroform	17.8	3.1	5.7
Diethyl ether	14.5	2.9	5.1
N,*N*-Dimethyl formamide	17.4	13.7	11.3
Ethanol	15.8	8.8	19.4
Ethyl acetate	15.8	5.3	7.2
n-Hexane	14.9	0.0	0.0
Tetrahydrofuran	16.8	5.7	8.0

3.3.2 Computational Polarity Modelling

To overcome the limitations of the solubility parameters, chemists can use more advanced modelling methods that require no experimental data to predict solvent polarity and consequently solubility, particularly the COnductor-like Screening MOdel for Real Solvents approach (COSMO-RS).[31] Firstly, the solute molecule is considered as embedded in a cavity that is surrounded by a virtual medium (COSMO). Secondly, the energies of interaction between molecules are quantified with statistical thermodynamics. These calculations allow for several representations of a molecule, in which it can be characterised as a substrate or as a solvent to provide solubility predictions. A medicinal chemist can use the COSMO-RS approach rather easily thanks to the user friendly *COSMOtherm* computer program developed by Klamt.[32]

The solvent characteristics generated by COSMO-RS can be used to compare solvents in order to identify promising 'drop-in' green replacements, *i.e.* solvents that can be used directly as substitutes. A sophisticated panorama of currently available bio-based solvents has been obtained through the COSMO-RS approach (Figure 3.2).[33,34] This representation permits a comparison between classical organic solvents and bio-based solvents. Unlike traditional organic solvents, bio-based solvents are poorly distributed throughout the available polarity space. Most of the bio-based solvents are either modestly dipolar and aprotic (*e.g.* 2-MeTHF, fatty acid methyl esters), or amphiprotic (*e.g.* ethanol, acetic acid, glycerol). Only a few have low polarities (*e.g.* limonene and the other terpenes). There is a real lack of bio-based solvents occupying the regions of polarity where classical organic solvents such as NMP or DMF can be found.

The chart in Figure 3.2 is comparable to equivalent graphs of the HSPs, but being purely predictive it can be applied to any bio-based solvent. The closer two solvents are spatially, the more similar they are likely to be in terms of polarity. If a chemist wants to find an alternative to a hazardous solvent, they only need to identify its position among this panorama and

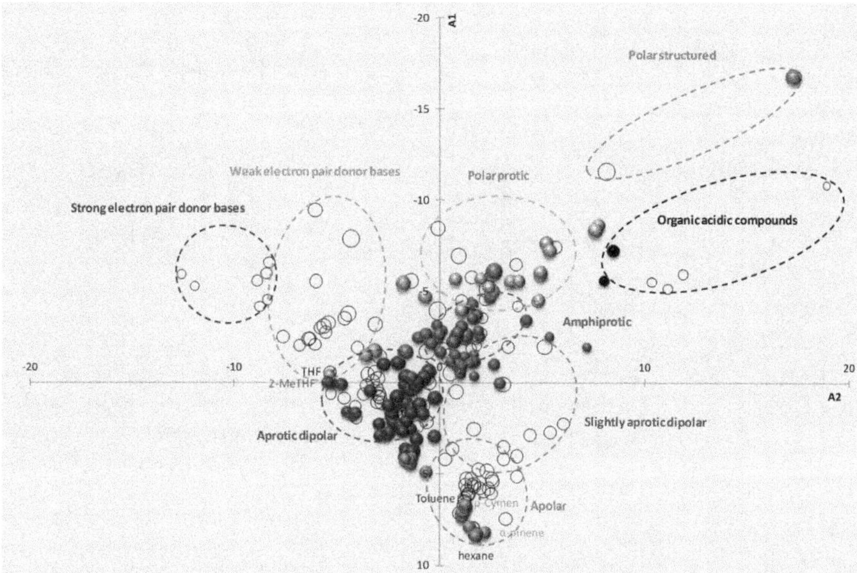

Figure 3.2 The 2D representation of a panorama of green solvents (coloured circles) among classical organic solvents (unshaded circles).
Adapted from Ref. 34 with permission from The Royal Society of Chemistry.

search for similar but renewable solvents. If several bio-based solvents happen to be similar to a hazardous solvent under scrutiny, the solvent candidates can be distinguished by predicting the solubility limit of a relevant substrate. This can be done using the COSMO-RS approach without any experimental data and with sufficient accuracy. One should note that this modelling approach highlights bio-based solvents with which to replace hazardous solvents only on the basis of solubilising performance. In reality, other constraints have to be considered too, such as volatility, viscosity and toxicity.

3.3.3 Linear Solvation Energy Relationships

In most instances, the components of a reaction will be soluble in a large number of solvents, and the main concern is usually not in what solvent to dissolve the reactants, but rather what solvent will promote the reaction and provide the most product in the shortest period of time. Polarity parameters based on the UV absorbance of solvatochromic dyes are suitable for correlations with kinetics and equilibria. Because the wavelength of the dye absorption is determined by the relative energy difference between the ground state and the excited electronic state of the probe molecule, it mirrors the variable energy levels of reaction components and intermediates in solution (Figure 3.3). Although several polarity scales of this type exist, the most reliable and broadly applicable are the Kamlet–Abboud–Taft

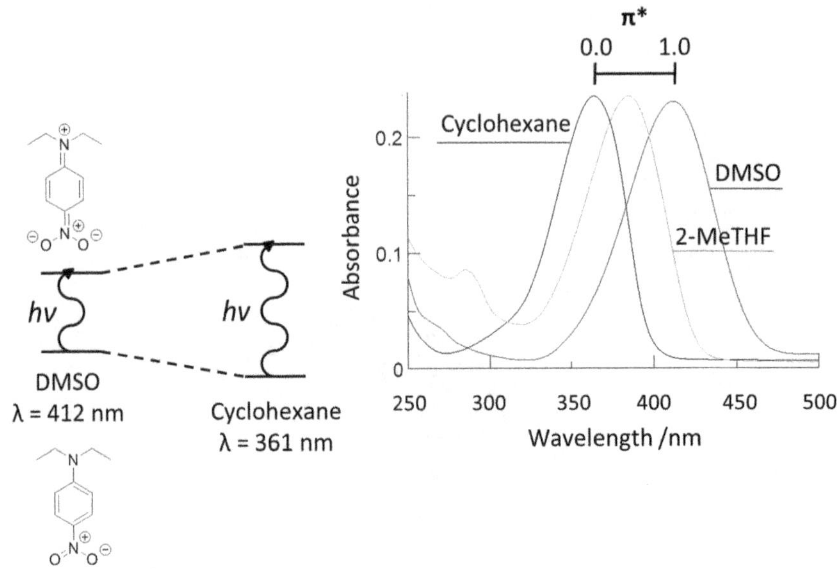

Figure 3.3 A schematic of the solvent effect on the electronic transition of *N,N*-diethyl-4-nitroaniline (left) and how its UV-spectra in different solvents translates to the π^* scale of dipolarity/polarisability (right).

solvatochromic parameters.[35] Derivable from the UV spectra of Reichardt's betaine dye, 4-nitroaniline, and *N,N*-diethyl-4-nitroaniline, the Kamlet–Abboud–Taft solvatochromic parameters manifest themselves as scales of hydrogen bond donating ability (α), hydrogen bond accepting ability (β), and general non-specific effects comprising of solvent dipolarity and polarisability (π^*).[21,23,35] For example, 2-MeTHF is aprotic ($\alpha = 0.00$), reasonably basic ($\beta = 0.57$) with a modest dipolarity ($\pi^* = 0.51$).

The three scales devised by Kamlet, Abboud and Taft have been used many times to formulate relationships between reaction rate constants and solvent polarity. These are known as linear solvation energy relationships (LSERs). The rate of amide formation for example, the most common single reaction in medicinal chemistry,[36] is inversely proportional to β for entropic reasons (Figure 3.4).[21] Limonene and its derivative *p*-cymene were thus justified as excellent options for a renewable amidation solvent, not only in terms of performance but also because they are produced from a renewable feedstock. Other solvents are less suitable according to their solvatochromic polarity parameters (Table 3.3). As hydrocarbons, some aquatic toxicity concerns surround the use of limonene and *p*-cymene, but ideally these would be minimised with recycling.

At the other end of the polarity scale, some progress is now being made in efforts to replace the reprotoxic dipolar aprotic amide solvents that we previously highlighted as lacking suitable alternatives. Using the LSER approach, cyclic carbonates have been shown to promote the rate of Heck

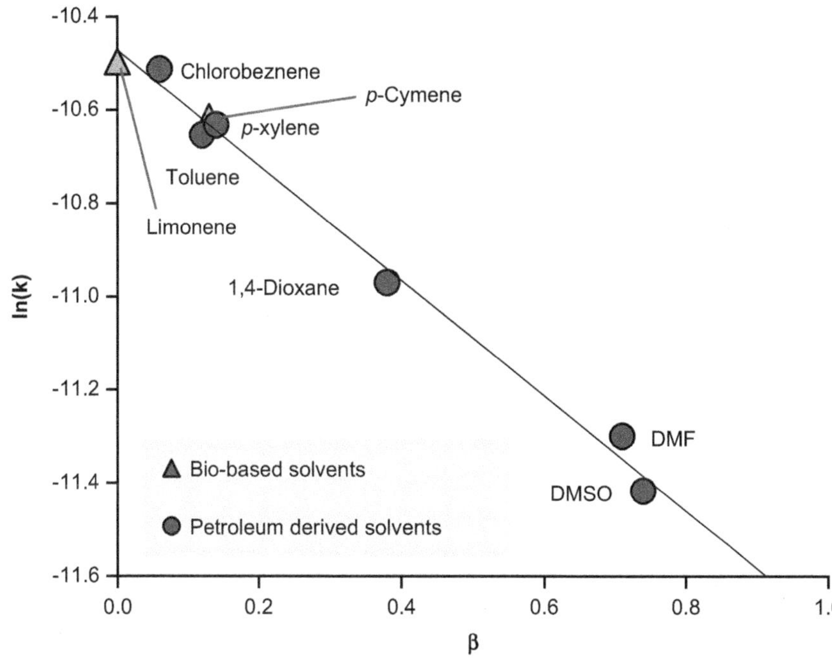

Figure 3.4 The relationship between the rate of amidation to give *N*-benzyl-4-phenylbutanamide and β.

Table 3.3 The polarity of solvents as derived from solvatochromic dyes.[35]

Solvent	α	β	π^*
Acetic acid	0.71	0.40	0.60
Acetone	0.00	0.51	0.70
Acetonitrile	0.35	0.37	0.80
Benzene	0.00	0.10	0.59
1-Butanol	0.73	0.85	0.61
Chloroform	0.20	0.10	0.58
Diethyl ether	0.00	0.51	0.28
N,*N*-Dimethyl formamide	0.00	0.71	0.88
Ethanol	0.83	0.77	0.62
Ethyl acetate	0.00	0.48	0.54
n-Hexane	0.00	0.00	−0.05
Tetrahydrofuran	0.00	0.55	0.58

cross-coupling (Figure 3.5).[37] This class of reaction is reliant on highly polar solvents. Cyclic carbonates can be produced from carbon dioxide.[38] In the future, diols produced by fermentation could serve as precursors to propylene carbonate and ethylene carbonate. Cyrene is another neoteric bio-based solvent that has been demonstrated to act as a replacement for NMP.[39] It is produced by the pyrolysis of waste cellulose (Scheme 3.3).

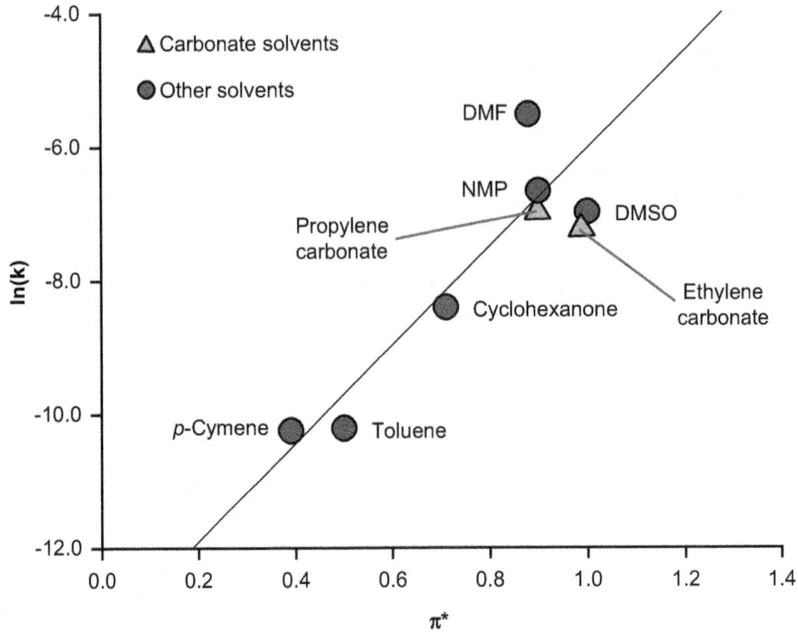

Figure 3.5 The rate of a Heck reaction between iodobenzene and methyl acrylate correlated with π^*.

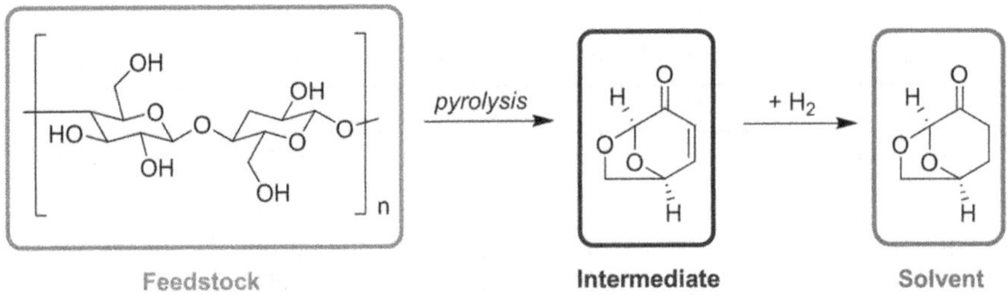

Scheme 3.3 The synthesis of Cyrene (dihydrolevoglucosenone).

3.4 Chapter Summary

A greater awareness of the origin of solvents, and why they are suitable for some reactions but not others, will be an essential quality of the future medicinal chemist. They will also need to embrace unconventional solvents in order to continue practicing the reactions familiar to medicinal chemistry.[40,41] Computational tools such as COSMO-RS are available for solvent selection that can improve the understanding and uptake of bio-based solvents. For example, limonene and *p*-cymene are bio-based solvent substitutes for traditional hydrocarbon solvents.

Growing academic interest in the field of neoteric solvents is beginning to permeate into industry, with bio-based solvents well placed to enhance the sustainability of medicinal chemistry. How the benefits of solvent selection are perceived by medicinal chemists is important in order to expand uptake of bio-based solvents in the pharmaceutical industry. Ultimately, company policies calling for culture change with respect to solvent use are needed to initiate a real and far reaching change.

References

1. T. J. Farmer and M. Mascal, Platform Molecules, in *Introduction to Chemicals from Biomass*, ed. J. H. Clark and F. Deswarte, John Wiley and Sons, Chichester, 2nd edn, 2015, pp. 89–156.
2. T. C. Ezeji, N. Qureshi and H. P. Blaschek, *Curr. Opin. Biotechnol.*, 2007, **18**, 220.
3. R. K. Henderson, C. Jiménez-González, D. J. C. Constable, S. R. Alston, G. A. A. Inglis, G. Fisher, J. Sherwood, S. P. Binks and A. D. Curzons, *Green Chem.*, 2011, **13**, 854.
4. Y. Y. Gorbanev, S. Kegnæs, C. W. Hanning, T. W. Hansen and A. Riisager, *ACS Catal.*, 2012, **2**, 604.
5. E. C. Corker, U. V. Mentzel, J. Mielby, A. Riisager and R. Fehrmann, *Green Chem.*, 2013, **15**, 928.
6. SEKAB ethyl acetate, http://www.sekab.com/chemistry/ethyl-acetate/ (accessed 22/06/2015).
7. C. S. M. Pereira, V. M. T. M. Silva and A. E. Rodrigues, *Green Chem.*, 2011, **13**, 2658.
8. T. Gerlach, F. Haese, A. Meier, J.-P. Melder and H. Rütter, *US Pat.* 7642382 B2, 2010.
9. A. Morschbacker, *J. Macromol. Sci., Polym. Rev.*, 2009, **49**, 79.
10. Regulation (EC) 1907/2006 of the European parliament and of the council concerning the Registration, Evaluation, Authorisation and Restriction of Chemicals (REACH), http://eur-lex.europa.eu/LexUriServ/LexUriServ.do?uri=OJ:L:2007:136:0003:0280:en:PDF (accessed on 16/06/2015).
11. J. H. Clark, L. A. Pfaltzgraff, V. L. Budarin, A. J. Hunt, M. Gronnow, A. S. Matharu, D. J. Macquarrie and J. R. Sherwood, *Pure Appl. Chem.*, 2013, **85**, 1625.
12. D. Tilman, R. Socolow, J. A. Foley, J. Hill, E. Larson, L. Lynd, S. Pacala, J. Reilly, T. Searchinger, C. Somerville and R. Williams, *Science*, 2009, **325**, 270.
13. C. Zhang, H. Su, Z. Wang, T. Tan and P. Qin, *Appl. Biochem. Biotechnol.*, 2015, **175**, 3901.
14. A. J. Hunt, E. H. K. Sin, R. Marriott and J. H. Clark, *ChemSusChem*, 2010, **3**, 306.
15. P. E. Savage, S. Gopalan, T. I. Mizan, C. J. Martino and E. E. Brock, *AIChE*, 1995, **41**, 1723.
16. G. Guiochon and A. Tarafder, *J. Chromatogr. A*, 2011, **1218**, 1037.

17. F. M. Kerton and R. Marriott, Green Solvents Legislation and Certification, in *Alternative Solvents for Green Chemistry*, RSC, Cambridge, 2nd edn, 2013, pp. 31–50.
18. M. Arshadi, A. J. Hunt and J. H. Clark, *RSC Adv.*, 2012, **2**, 1806.
19. D. F. Aycock, *Org. Process Res. Dev.*, 2007, **11**, 156.
20. A. Kadam, M. Nguygen, M. Kopach, P. Richardson, F. Gallou, Z. K. Wan and W. Zhang, *Green Chem.*, 2013, **15**, 1880.
21. J. H. Clark, D. J. Macquarrie and J. Sherwood, *Green Chem.*, 2012, **14**, 90.
22. J. H. Clark, E. M. Fitzpatrick, D. J. Macquarrie, L. A. Pfaltzgraff and J. Sherwood, *Catal. Today*, 2012, **190**, 144.
23. J. H. Clark, D. J. Macquarrie and J. Sherwood, *Chem. – Eur. J.*, 2013, **19**, 5174.
24. R. Ciriminna, C. D. Pina, M. Rossi and M. Pagliaro, *Eur. J. Lipid Sci. Technol.*, 2014, **116**, 1432.
25. Y. Gu and F. Jérôme, *Green Chem.*, 2010, **12**, 1127.
26. P. J. Dunn, *Green Chem.*, 2013, **15**, 3099.
27. K. Grodowska and A. Parczewski, *Acta Pol. Pharm.*, 2010, **67**, 3.
28. D. J. C. Constable, C. Jiménez-González and R. K. Henderson, *Org. Process Res. Dev.*, 2007, **11**, 133.
29. C. Reichardt, *Solvents and Solvent Effects in Organic Chemistry*, Wiley-VCH, Weinheim, 3rd edn, 2003.
30. C. M. Hansen, *Hansen Solubility Parameters: A User's Handbook*, Taylor & Francis, Boca Raton, 2nd edn, 2007.
31. A. Klamt, *Cosmo-RS: From Quantum Chemistry to Fluid Phase Thermodynamics and Drug Design*, Elsevier Science, Amsterdam, 1st edn, 2005.
32. COSMOlogic software product COSMO-RS, http://www.cosmologic.de/theory/cosmo-rs.html (accessed 22/06/2015).
33. M. Durand, V. Molinier, W. Kunz and J. M. Aubry, *Chem. – Eur. J.*, 2011, **17**, 5155.
34. L. Moity, M. Durand, A. Benazzouz, C. Pierlot, V. Molinier and J. M. Aubry, *Green Chem.*, 2012, **14**, 1132.
35. P. G. Jessop, D. A. Jessop, D. Fu and L. Phan, *Green Chem.*, 2012, **14**, 1245.
36. S. D. Roughley and A. M. Jordan, *J. Med. Chem.*, 2011, **54**, 3451.
37. H. L. Parker, J. Sherwood, A. J. Hunt and J. H. Clark, *ACS Sustainable Chem. Eng.*, 2014, **2**, 1739.
38. M. North, R. Pasquale and C. Young, *Green Chem.*, 2010, **12**, 1514.
39. J. Sherwood, M. De bruyn, A. Constantinou, L. Moity, C. R. McElroy, T. J. Farmer, T. Duncan, W. Raverty, A. J. Hunt and J. H. Clark, *Chem. Commun.*, 2014, **50**, 9650.
40. S. W. Breeden, J. H. Clark, D. J. Macquarrie and J. Sherwood, Green Solvents, in *Green Techniques for Organic Synthesis and Medicinal Chemistry*, ed. W. Zhang and B. W. Cue Jr., John Wiley and Sons, Chichester, 2012, pp. 243–261.
41. Y. Gu and F. Jérôme, *Chem. Soc. Rev.*, 2013, **42**, 9550.

CHAPTER 4

Beyond Mass-based Metrics: Evaluating the Greenness of Your Reaction

LOUISE SUMMERTON* AND ANDRI CONSTANDINOU

Green Chemistry Centre of Excellence, Department of Chemistry, University of York, York YO10 5DD, UK
*Email: louise.summerton@york.ac.uk

4.1 Introduction

Improving the sustainability of research does not happen overnight, but is an iterative process whereby a series of improvements will be made over a period of time. Metrics should be used in order to monitor, measure and evaluate this. To strike a balance between obtaining a meaningful assessment and reducing time spent on the process, the metrics applied should be simple but not simplistic. Focussing on mass-based metrics alone will not suffice in terms of obtaining a true picture of the greenness of a reaction. This view is supported by the review carried out by Scientific Update with 24 chemical companies (including 17 pharmaceutical companies), which found that most pharmaceutical companies use more than one metric in-house, and some use as many as ten different metrics to assess greenness.[1] Adopting a variety of metrics, incorporating life cycle considerations, is the ideal way to measure 'greenness' of a process.[2] The risk of focussing on only one parameter is that when making an improvement in one aspect of the reaction/methodology this could have detrimental effects elsewhere. It is

RSC Green Chemistry No. 46
Green and Sustainable Medicinal Chemistry: Methods, Tools and Strategies for the 21st Century Pharmaceutical Industry
Edited by Louise Summerton, Helen F. Sneddon, Leonie C. Jones and James H. Clark

Published by the Royal Society of Chemistry, www.rsc.org

also possible to simply select the metrics that would make the research 'look good' without genuinely assessing its true 'green credentials'.

4.2 Boundaries of Metrics Assessment

When assessing the greenness of a reaction, it is essential to incorporate isolation/purification methodologies and downstream processing (DSP) in order to gain a holistic overview. This is to ensure potential environmental hot spots are not ignored downstream from the actual synthetic step.[3] In fact, a recent study performed by GlaxoSmithKline on in-house production found that less than 25% of its unit operations are related to reactions, with the rest being work-up procedures, and that 40–90% of the calculated process mass intensity (PMI) for a pharmaceutical process is attributable to separation steps.[4]

Significant life cycle burdens are often generated during the downstream processes for bio-reactions.[5] The product streams from bio-reactions are typically dilute aqueous solutions, which can bring specific problems that need to be addressed from both an environmental and process optimisation standpoint, for example large amounts of organic solvents may be used for extraction.[6]

Another downstream issue that should be considered is that of recycling and recovery (*e.g.* of solvents and catalysts). This is important and can have a positive effect on the environmental impact and hence the metrics of a given process. Studies have been carried out to determine whether solvent recovery is viable depending upon factors such as the energy required to achieve this.[7] ETH Zurich has also developed an LCA based tool that allows the environmental impact of waste-solvent treatment to be quantified.[8]

Recommendation: Don't look at your reaction step in isolation as environmental issues can be hidden in upstream and downstream processes. In particular, consider what methods you are using for isolation and work-up procedures.

4.3 Tracking Improvements

The environmental impacts of the synthesis of pharmaceutical products/intermediates are influenced by both the complexity of the chemistries employed in the route and the inherent molecular complexity of the API or intermediate itself.[9] Therefore, it is not always appropriate to monitor improvements by setting metric 'targets' for a reaction in terms of minima/maxima values, as this does not allow for the molecular complexity of the targets.

However, options are available to adjust for complexity, for example, a PMI calculation has been established by Lilly that sets targets for the PMI of API synthesis based on a correlation between the number of chiral centres, the number of heteroatoms and the fraction of aromatics in the target molecule.[10]

Another option for tracking improvements is the use of EcoScale, whereby reactions are scored on parameters that favour high-yielding, low-cost and safe reaction conditions with an easy purification.[11,12] Although this provides certain benefits, there is a danger that simply assigning an overall 'number' or score to a reaction or methodology may not be the most meaningful way of assessing its 'greenness' as it does not allow one to identify areas of concern, or indeed aspects where improvements are being made.

A modified version of the EcoScale has been adopted by Boehringer Ingelheim to define a 'good chemical manufacturing process', in which a reaction step is evaluated based on eight criteria (yield; quality; work-up and purification; equipment; reaction time; reaction temperature; raw materials; and EHS).[13] The majority of the categories are divided into two or three criterion, for which high, mid-range or low scores are attributed.

Recently, Roschangar *et al.* published their Green Aspiration Level™ concept, a process performance metric for the pharmaceutical industry aimed at assessing the relative greenness of a given process with relation to waste, whilst at the same time factoring in the complexity of the process.[14]

Recommendation: Improvements in metrics are easier to follow if data can be separated out, for example breaking down mass inputs into categories for process mass intensity (PMI) calculation and hence obtaining a PMI value for solvents, PMI for water and PMI for all other reagents, as well as a total figure. This can also help identify environmental 'hot spots' in terms of the largest contributor to the overall figure.

4.4 Factors to Take into Consideration

When a medicinal chemist is evaluating the greenness of their reaction there are a number of considerations that can be taken into account, namely:

- Efficiency
- Waste
- Solvents
- Catalyst/enzyme
- Elemental sustainability
- Energy
- Health and safety
- Chemicals of environmental concern

These factors are discussed in more detail below, alongside some straightforward recommendations for where improvements can potentially be made. The CHEM21 (Chemical Manufacturing Methods for the 21st Century Pharmaceutical Industries) consortium has been working on the development of a unified sustainability metrics toolkit that combines all of the above factors to allow members to evaluate and verify the 'green credentials' of their research.[15,16] The toolkit is described in more detail later in this chapter.

Details of the metrics identified, and their relative advantages and disadvantages for measuring 'greenness' are also discussed in Table 4.1.

Table 4.1 Details of some available metrics and their relative advantages and disadvantages for measuring 'greenness'.

Metric	Equation	Discussion
Yield	$\%\,\text{Yield} = \dfrac{\text{moles of product}}{\text{moles of limiting reactant}} \times 100$	The most well-known and commonly used metric in chemistry is that of yield. It expresses the number of molecules of limiting reactant that are converted to product compared to the overall number present at the beginning of the reaction.
Atom Economy (AE)	$\dfrac{\text{MW of product(s)}}{\text{MW of all reactants}} \times 100\,\%$	Atom economy is a theoretical number that is based on a chemical yield of 100% of theoretical and does not take into account the reactants stoichiometry.[36] It does not include inorganic reagents and solvents and ignores molar excess of reactants.[37]
Reaction Mass Efficiency (RME)	$\text{RME} = \dfrac{\text{mass of isolated product}}{\text{total mass of reactants}} \times 100$	RME is an improved form of atom economy that takes the yield and the amount of reactants into account.[36] However, compared with AE it is more time consuming, and another issue is the potential difficultly of differentiating between what is classed as a reactant and what is classed as an auxiliary. It is more comprehensive than atom economy as it also incorporates the effect of yield and reagent excesses. Another benefit of the RME calculation is that it can be varied by a so-called 'material recovery factor' to account for the downstream purification methods adopted, *e.g.* whether or not the catalyst/reaction solvent/work-up and purification materials are recovered.[38]

Mass Intensity (MI)/ Process Mass Intensity (PMI)	$\frac{\text{total mass in a process or process step}}{\text{mass of product}}$	Mass intensity takes into account the yield, reaction stoichiometry, solvents, and reagents in a reaction mixture as well as in the reaction workup and purification steps. It does not include any process water used. For a multiple step process, process mass intensity is used instead where the mass inputs include everything from step one to the last step. The ideal PMI is 1 when no waste is produced and all materials are integrated into the product. By simply manipulating the top line of the equation to show total mass of: reaction, solvents, reagents, work up *etc.*, it's possible to break down a process into these individual components and generate mass intensities for each. This is invaluable for the researcher as not only is there an overall value against which to evaluate any new methodology against the state of the art, but the breakdown highlights in which areas a particular process scores well and in which areas work needs to be focused on to bring about improvements.
E-factor	$E_m = \frac{1}{\text{RME}} - 1$	*E*-factor is the actual amount of waste produced in a process. It takes the chemical yield into account and includes all reagents, solvent losses, all process aids, excluding water however because it would lead to exceptionally high *E*-factor figures, which eventually make the comparisons of processes difficult. The definition of total waste is not entirely clear, and boundaries must be drawn to decide whether factors such as emissions, energy use and solvent are accounted for.[37]

4.4.1 Efficiency

Metrics that determine efficiency are generally based on mass inputs, and are relatively easy to calculate and interpret. In a survey of 24 chemical companies (including 17 pharmaceutical companies) the most popular metric was found to be process mass intensity (PMI; 67% of companies), followed by *E*-factor (48% of companies).[1] PMI has been adopted by the ACS GCI Pharmaceutical Roundtable (ACS GCI PR) as the 'key, high-level metric for evaluating and benchmarking progress towards more sustainable manufacturing'.[2] Efficiency-based metrics tend to be favoured over waste metrics in a business context as they communicate sustainability in terms of adding value by increasing productivity instead of managing the costs of reducing waste.[2]

Recommendation: Improvements in reaction efficiencies are possible by, for example, designing reactions not to use auxiliaries, choosing leaving groups that are as small as possible and considering the stoichiometric loading of reagents.

4.4.2 Waste

Quantities of waste produced can be calculated readily by analysing mass outputs using, for example, *E*-factor calculations. However, this does not give insight into the nature of the waste itself or other life cycle considerations, such as how the waste is disposed of. Problems can arise with mixed organic–aqueous waste streams, which may require further unit operations to separate the streams before sending to waste water treatment operations, or alternatively waste may be incinerated.[17]

Recommendation: In addition to performing E factor or other waste calculations, take time to consider the nature of the waste for example from a health and safety point of view.

4.4.3 Solvents

From a number of cradle-to-gate life cycle studies it is generally recognised that the major contributor to the life cycle impacts of API synthesis is due to solvents. A large number of solvent selection guides are available in the literature and have been developed in-house by a number of pharmaceutical companies, including GlaxoSmithKline, Pfizer, Sanofi and Astra Zeneca.[18–20] The ACS GCI PR has also produced an industry-wide solvent selection guide, which is available as a free application for mobile phones.[21] (See Chapters 2 and 3 for more detail on solvent selection and tools.)

When considering the solvents used in a synthetic pathway to an API or intermediate, one quick analysis that may be performed is simply to look at the number and types of solvents used. Counting the number of solvent swaps may identify opportunities where steps could be carried out in succession using the same solvent, without the need to isolate the product (telescoping reactions).[22]

Recommendation: Alongside the use of solvent selection guides to try to identify as green a solvent for your reaction as possible, consider whether reaction steps could be carried out in the same solvent and hence opening up the possibility for telescoping reactions and easier solvent recycling.

4.4.4 Catalyst/Enzyme

The use of catalytic as opposed to stoichiometric reagents is obviously favoured from a green chemistry perspective.[23] However, a question that must be asked is 'how green is the catalyst itself?' The preparation of a catalyst can have associated environmental issues. For example, a cradle-to-gate life cycle study of enzymes for pharmaceutical applications concluded that the production of immobilised enzymes is generally energy intensive.[5] To ignore catalyst preparation from a metrics perspective could be to ignore significant environmental issues, such as the use of particularly toxic reagents. Other key considerations regarding the use of a catalyst include turnover number (TON), space time yield (important when planning scale-up), recoverability/reuse and the use of critical elements.

Recommendation: Consider whether your catalyst is recoverable, for example, could it be supported/immobilised or could your reaction be performed in a biphasic solvent system?

4.4.5 Elemental Sustainability

Raw materials are classed as critical when the risks for supply shortage and their impacts on the economy are higher when compared with most other raw materials. Platinum group metals (PGMs; platinum, palladium, iridium, rhodium, ruthenium and osmium) are one of the groups of materials that have been classed as critical at EU level (Brussels MEMO/10/263). Metals from this group are widely employed as catalysts in the synthesis of pharmaceutical products. Currently there is no actual metric available that accounts for critical elements, although it is possible to refer to data sources such as the British Geological Risk List, which indicates the relative risk to the supply of a number of elements based on factors affecting their availability.[24] (See Chapter 5 for more detail on elemental sustainability.)

Recommendation: Consider substituting a critical element for a less critical element, or if this is not feasible, consider whether it could be recovered.

4.4.6 Energy

Energy consumption is not straightforward to measure on a lab scale, as even using an energy meter there will be discrepancies between different pieces of equipment, their efficiencies, heating rates *etc.*, making comparisons difficult. One way of factoring in energy considerations whilst overcoming these barriers is the method used by Boehringer Ingelheim mentioned earlier, where energy consumption is correlated with temperature ranges *i.e.* over 150 °C and below −10 °C are penalised in terms of their overall score.[13]

Recommendation: If your reaction falls outside a standard temperature range, consider whether your reaction could be performed under milder conditions. Energy savings can be made simply by running a reaction slightly below reflux.

4.4.7 Health and Safety

A gap in the quantitative metrics available to measure the 'greenness' of a reaction is the actual nature of the chemicals used, *i.e.* their inherent toxicity or hazards associated with them and the reactions themselves (*e.g.* highly exothermic reactions). Material safety data sheets (MSDS) are obviously a reliable source of hazard information for commercially available materials. One method for measuring occupational exposure, hazard and risk is *via* hazard banding.[17] The Swiss Federal Institute of Technology, Zurich (ETH Zurich) has developed a free-to-use EHS assessment tool (Excel-based) that allows users to identify safety, health and environmental issues of a chemical process, find their source and estimate their magnitude.[25] The tool was specifically developed for the production of speciality chemicals (*i.e.* pharmaceuticals, agro chemicals and fine chemicals).[26] Chemicals are searched by substance name, CAS number or UN number and up to 10 chemicals can be included at one time. Eleven categories are assessed by the tool. For safety these are release potential; fire/explosion; and reaction/decomposition, for health they are acute toxicity; irritation and chronic toxicity, and for environment they are persistency; air hazard; and water hazard. The results can be displayed as graphs or tables, either for the overall EHS result or the user can look at each individual category separately.

Recommendation: If any of the compounds used in your reaction carry a substantial physical, health or environmental risk, consider whether they can be substituted by a more benign replacement.

4.4.8 Chemicals of Environmental Concern

Acute and chronic toxicity data of chemicals is being gathered in response to legislation such as REACH. We must be aware of the potential for chemicals to appear on regulatory lists and avoid their use wherever possible. Two interesting sources of data on such issues have been developed by an NGO based in Sweden—ChemSec, the International Chemical Secretariat. ChemSec is a highly regarded organisation with many major corporate supporters. The SIN List (Substitute it Now) consists of 378 chemicals that ChemSec has identified as substances of very high concern based on the criteria established by the EU chemical regulation, REACH.[27] While the SIN List is unofficial, a number of multinational companies have chosen to use it for example in instructing their suppliers in which chemicals to avoid. SUBSPORT, the Substitution Support Portal, acts as a database/search engine on hazardous substances that are legally or voluntarily restricted or subjects of public debates, and acts as a substitution guide.[28] (See Chapter 1.)

Recommendation: If any of the compounds used in your reaction have been identified as a chemical of concern, e.g. by REACH, consider whether they can be substituted by a more benign replacement.

4.5 Wider Considerations

4.5.1 Renewability

The use of renewables, *i.e.* bio-derived organic reagents and solvents, is an important step towards improving the sustainability of a given process, but does of course need to be assessed within a life cycle context, as potential trade-offs are inevitable in terms of, for example, land use. The EU has started a major programme to introduce bio-based standards and the associated technical committee (TC411) is examining standards for especially important products including solvents. Standardisation is necessary to improve consistency of product quality, assist in the harmonisation of the industry and improve consumer confidence.[29] (See Chapter 1 for more detail on renewables.)

4.5.2 Life Cycle Assessment

To gain a comprehensive view of the true environmental impacts of products and processes requires life cycle assessment (LCA) studies to be performed. Parameters that are measured as part of a LCA include: total cradle mass (amount of materials taken from the earth), energy requirements, greenhouse gas emissions (GHG), photochemical ozone creation (POCP), eutrophication, acidification, and total organic carbon (TOC).[30] Full LCA is extremely time-consuming and life cycle inventory (LCI) data is often difficult to acquire, in particular for bioprocesses in terms of substrates and enzymes.[6]

There is a lack of industry-wide streamlined LCA tools that are easy to use, consistent and transparent.[4] Two such tools, GSK's FLASC™ tool and the ACS GCI PR PMI-LCA tool, are not yet publically available.[9,31] Other free-to-use LCA tools are however available, such as CCaLC and Finechem. The CCaLC Carbon Footprinting tools were developed as part of a collaborative project led by the University of Manchester (funded by EPSRC, NERC and the Carbon Trust) to provide quick and easy estimations of the life cycle greenhouse gas emissions along whole supply chains.[32] The tools follow internationally accepted life cycle methodology as defined by ISO 14044. CCaLC also calculates additional environmental impacts to demonstrate how they may be affected by changes in the carbon footprint, including acidification, eutrophication, ozone layer depletion, photochemical smog and human toxicity.[32] ETH Zurich has also developed Finechem, a free of charge software tool that can approximate Cumulative Energy Demand

(CED) and Global Warming Potential (GWP) values of petrochemical based feedstocks from their molecular structures.[33] However, it is worth noting that products that include biosynthesis steps in their production were not included in the data upon which the tool was modelled.[34]

4.6 CHEM21 Metrics Toolkit

The CHEM21 project has developed a unified metrics toolkit to comprehensively evaluate the sustainability of chemical and bio-chemical reactions and encourage continuous improvement.[16] To ensure a genuinely holistic approach, the toolkit uses a blend of both qualitative and quantitative criteria to assess how green a reaction is, as well as considering both upstream and downstream of the reaction itself. By examining a broad range of criteria at the same time, the toolkit makes sure that by ameliorating one issue others do not occur elsewhere in the process.

The interpretation of the metrics assessment from the toolkit highlights where research is performing well in terms of its 'greenness' and suggests areas where improvements could potentially be made. The toolkit can also be used to assess state-of-the-art reactions/routes to target compounds to provide a benchmark against which to compare results.

The toolkit is structured so that it becomes increasingly rigorous in its assessment *via* a series of passes, going from small scale bench top research right through to industrial scale (see Figure 4.1).[16]

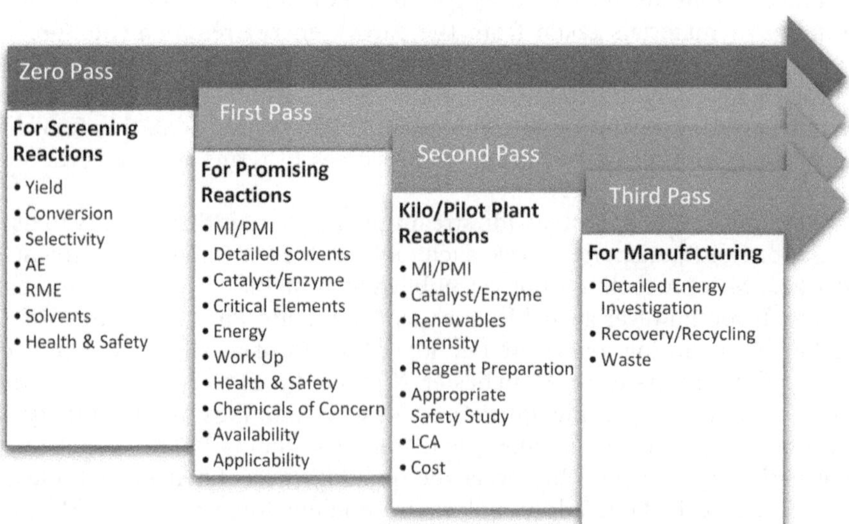

Figure 4.1 Structure of the CHEM21 metrics toolkit showing the parameters covered at each pass (Reproduced from ref. 16 – Published by The Royal Society of Chemistry).

The main goal of the CHEM21 Metrics Toolkit is to encourage critical thinking surrounding the sustainability and environmental acceptability of reactions, and ultimately make analysis of synthetic routes and the use of greener and more sustainable techniques part of everyday practice amongst the academic and industrial communities. To achieve this, the CHEM21 Metrics Toolkit has been made freely available in the form of a user-friendly Excel spreadsheet.[35]

4.7 Conclusion

Metrics will not provide a definitive answer to the question "is my reaction green?"—this is never a straightforward yes or no answer as there are numerous factors to take into consideration. An approach that encompasses more than mass-based metrics as described above, alongside a greater understanding of the issues surrounding sustainability, and the ability to intepret the metrics results should however provide evidence to suggest where improvements could be made, and where the methodology is performing well. It should also help answer probably the most important question, "is my reaction **greener**?" As with all things, a balance must be struck. So, if making a modification to, for example, the temperature of your reaction to milder conditions, leads to a detrimental effect on other metrics, such as mass efficiencies, then it is likely that this change is not beneficial in terms of overall greenness. Hence, once again this demonstrates the need for a more holistic point of view, incorporating a number of considerations into the decision-making process.

4.8 Summary Points

- The use of mass-based metrics in isolation does not give a true picture of the greenness of a reaction
- Metrics selected should not simply focus on a single reaction step in isolation, but also consider upstream and downstream options.
- Issues such as elemental sustainability and health and safety do not have simple metrics associated with them, and require the chemist themselves to be aware of the 'bigger picture' in terms of the issues surrounding their use, and what measures can be taken to either substitute them or mitigate their impact.

References

1. W. J. W. Watson, *Green Chem.*, 2012, **14**, 251.
2. C. Jimenez-Gonzalez, C. S. Ponder, Q. B. Broxterman and J. B. Manley, *Org. Process Res. Dev.*, 2011, **15**, 912.
3. J. H. Clark, S. W. Breeden and L. Summerton, *Future Med. Chem.*, 2009, **1**, 409.

4. C. Jiménez-González, P. Poechlauer, Q. B. Broxterman, B.-S. Yang, D. am Ende, J. Baird *et al.*, *Org. Process Res. Dev.*, 2011, **15**, 900.
5. S. Kim, C. Jiménez-González and B. Dale, *Int. J. Life Cycle Assess.*, 2009, **14**, 392.
6. C. Jiménez-González and J. M. Woodley, *Comput. Chem. Eng.*, 2010, **34**, 1009.
7. M. J. Raymond, C. S. Slater and M. J. Savelski, *Green Chem.*, 2010, **12**, 1826.
8. http://www.sust-chem.ethz.ch/tools/ecosolvent (last accessed July 2015).
9. A. Curzons, C. Jiménez-González, A. Duncan, D. C. Constable and V. Cunningham, *Int. J. Life Cycle Assess.*, 2007, **12**, 272.
10. D. P. Kjell, I. A. Watson, C. N. Wolfe and J. T. Spitler, *Org. Process Res. Dev.*, 2013, **17**, 169.
11. http://www.ecoscale.org (last accessed July 2015).
12. K. Van Aken, L. Strekowski and L. Patiny, *Beilstein J. Org. Chem.*, 2006, **2**, 3.
13. R. Dach, J. J. Song, F. Roschangar, W. Samstag and C. H. Senanayake, *Org. Process Res. Dev.*, 2012, **16**, 1697.
14. F. Roschangar, R. A. Sheldon and C. H. Senanayake, *Green Chem.*, 2015, **17**, 752.
15. http://www.chem21.eu/ (last accessed July 2015).
16. C. R. McElroy, A. Constantinou, L. C. Jones, L. Summerton and J. H. Clark, *Green Chem.*, 2015, **17**, 3111.
17. C. Jimenez-Gonzalez, D. J. C. Constable and C. S. Ponder, *Chem. Soc. Rev.*, 2012, **41**, 1485.
18. R. K. Henderson, C. Jimenez-Gonzalez, D. J. C. Constable, S. R. Alston, G. G. A. Inglis, G. Fisher *et al.*, *Green Chem.*, 2011, **13**, 854.
19. K. Alfonsi, J. Colberg, P. J. Dunn, T. Fevig, S. Jennings, T. A. Johnson *et al.*, *Green Chem.*, 2008, **10**, 31.
20. D. Prat, O. Pardigon, H.-W. Flemming, S. Letestu, V. Ducandas, P. Isnard *et al.*, *Org. Process Res. Dev.*, 2013, **17**, 1517.
21. https://itunes.apple.com/us/app/green-solvents/id446670983 (last accessed July 2015).
22. G. Assaf, G. Checksfield, D. Critcher, P. J. Dunn, S. Field, L. J. Harris *et al.*, *Green Chem.*, 2012, **14**, 123.
23. http://www.acs.org/content/acs/en/greenchemistry/what-is-green-chemistry/principles/12-principles-of-green-chemistry.html (last accessed July 2015).
24. https://www.bgs.ac.uk/downloads/start.cfm?id=2643 (last accessed July 2015).
25. http://www.sust-chem.ethz.ch/tools/ehs (last accessed July 2015).
26. G. Koller, U. Fischer and K. Hungerbühler, *Ind. Eng. Chem. Res.*, 2000, **39**, 960.
27. http://sinlist.org/ (last accessed July 2015).
28. http://www.subsport.eu/ (last accessed July 2015).
29. http://www.biobasedeconomy.eu/research/kbbpps/ (last accessed July 2015).

30. C. Jiménez-González, A. Curzons, D. C. Constable and V. Cunningham, *Int. J. Life Cycle Assess.*, 2004, **9**, 114.
31. C. Jiménez-González, C. Ollech, W. Pyrz, D. Hughes, Q. B. Broxterman and N. Bhathela, *Org. Process Res. Dev.*, 2013, **17**, 239.
32. http://www.ccalc.org.uk/ (last accessed July 2015).
33. http://www.sust-chem.ethz.ch/tools/finechem (last accessed July 2015).
34. G. Wernet, S. Papadokonstantakis, S. Hellweg and K. Hungerbuhler, *Green Chem.*, 2009, **11**, 1826.
35. http://www.rsc.org/suppdata/c5/gc/c5gc00340g/c5gc00340g1.xlsx (last accessed July 2015).
36. R. A. Sheldon, *Chem. Soc. Rev.*, 2012, **41**, 1437.
37. D. J. C. Constable, A. D. Curzons and V. L. Cunningham, *Green Chem.*, 2002, **4**, 521.
38. J. Andraos, *Green Chemistry Metrics: Measuring and Monitoring Sustainable Processes*, ed. A. Lapkin and D. J. C. Constable, Wiley-Blackwell, Oxford, vol. 1, 2008, p. 69.

CHAPTER 5

The Importance of Elemental Sustainability and Critical Element Recovery for the Pharmaceutical Industry

ANDREW J. HUNT

Green Chemistry Centre of Excellence, Department of Chemistry, University of York, York YO10 5DD, UK
Email: andrew.hunt@york.ac.uk

5.1 Elemental Sustainability

5.1.1 Why is Elemental Sustainability Important?

Today the pharmaceutical industry is still reliant on the oil industry to supply a significant proportion of the organic feedstocks. The demand for oil is still increasing, even after reaching peak oil production in 2006. In recent years, both the public and industry alike have become aware of the need to move towards being carbon neutral and reduce our reliance on fossil resources. This has resulted in a strong drive to move to the use of renewable materials and feedstocks. However, we are still far from being neutral in the use of all elements of the periodic table. This is an important issue that we must all address! The growing need to find sustainable solutions to alleviate climate change and our dependence on fossil fuel resources has led to the

RSC Green Chemistry No. 46
Green and Sustainable Medicinal Chemistry: Methods, Tools and Strategies for the 21st Century Pharmaceutical Industry
Edited by Louise Summerton, Helen F. Sneddon, Leonie C. Jones and James H. Clark

Published by the Royal Society of Chemistry, www.rsc.org

development of "low carbon technologies".[1] Many of these technologies contain a significant number of elements in their construction. Coupled with the increased production of and demand for consumer products in recent years (including electronics, cars, jewellery, food and pharmaceuticals, to name but a few), concerns over the security of future elemental accessibility and supply have increased. Frequently, rare and precious elements are incorporated into low carbon technologies or utilized in the production of consumer products (catalysts for pharmaceutical production), thus spreading these elements throughout our environment.[2,3] Many traditional sources of elements are in decline, creating further challenges in elemental extraction, manufacture, use and recovery, which can only be tackled through a multidisciplinary approach (Figure 5.1).[1,3,4] Currently, the vast majority of elements have low end-of-life recycling rates. Instead, waste materials are dispersed throughout the environment in low concentrations, making them very difficult to recover in the future.[5,6]

By increasing recycling rates, individual companies, industries or even nations can move towards a circular economy and thereby reduce their reliance on element production whilst also generating little, or ideally no, waste (Figure 5.1).[1]

Elemental sustainability is an integral part of a circular economy and is a concept that should be adopted by the pharmaceutical industry.[7] The key goal of elemental sustainability is to ensure that all elements within the periodic table are guaranteed for use by both current and future generations.[1] These significant challenges need to be addressed through a holistic approach to resource utilization. As such, it is essential to develop new sustainable routes and strategies for the recovery and reuse of these elements.

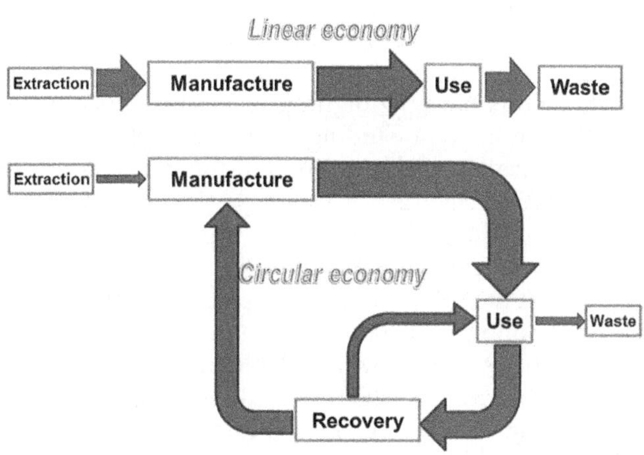

Figure 5.1 Illustration to highlight a simple material flow cycle: the areas that must be addressed to meet the challenges of elemental sustainability, which could lead to a circular economy.

5.1.2 What are Critical Elements?

The long-term security of critical elemental supply has become an important issue for many organisations (including the pharmaceutical industry). Elements with significant supply risk issues, which if restricted could harm the pharmaceutical industry, can be considered "critical".[7,8] However, for a true assessment of elemental sustainability, the environmental impact of extraction and use should also be taken into account. The extraction of elements from their ores is an energy intensive process and there is a significant environmental burden associated with these activities. The time between the discovery of new reserves and those elements entering the market can be significant, typically greater than 10 years. In the future, a greater proportion of elements may need to be sourced from waste, including landfill sites, waste electrical and electronic equipment, contaminated waters, industrial waste and mine tailings.

The definition of which elements are classed as critical can vary depending on the purpose of the assessment and needs of the organisation. Several national and international assessments of critical elements have been made, all of which concur that current and projected demands for that element will result in rapid depletion of known reserves and diffusion throughout the environment.[8–16] Hunt *et al.* collated the elements listed in these international assessments and highlighted those critical elements of global importance (Table 5.1).[1]

Population increases worldwide could have a significant impact on the pharmaceutical industry, as this will lead to greater demand for medicines. Critical elements frequently have a vital function (for example as catalysts) but often there is a low critical element content per unit, thus increasing the dispersion of such elements. Those elements classified as being of global importance (Table 5.1) typically have low production volumes and recycling rates compared to industrial metals.[5,6] Many of these critical elements are predominately produced as a by-product of mining or processing of base metals, resulting in difficult extractive metallurgies.[21] Several reports on critical elements reports have highlighted other elements of importance for the pharmaceutical industry, including helium and phosphorus. Helium is non-renewable and available from a limited number of sources, including natural gas deposits and the nuclear industry. The use of helium is vital for the pharmaceutical industry as it is extensively used in analytical techniques, including gas chromatography (GC) and nuclear magnetic resonance (NMR), for the determination of both structure and product purity. Approximately a third of the world's supply is current sourced from North America; however, it is predicted that these reserves will be depleted in the next 10 years. Other resources, such as fluorite (also known as fluorspar), an important source of fluoride, which can be used in the manufacture of fluorine-containing fine chemicals or pharmaceutical intermediates, have also been highlighted as critical by the EU.

Table 5.1 Critical selements of global importance. Adapted from ref. 29 with permission from The Royal Society of Chemistry.

Element	Resources remaining from traditional sources[a] (years)	Main geographical location of known resource	Current recycling rates (%)[b]
	Critical elements of global importance		
Rare earths	100+	China	<1
Gallium	5–50	China	<1
Indium	5–50	China	<1
Platinum group metals	5–50 (50–100 years for Pd)	South Africa	>50
Tantalum	50–100	Brazil	<1
Cobalt	50–100	Congo	>50
Niobium	50–100	Brazil	>50
Antimony	5–50	China	1–10
Beryllium	100+	USA	<1
Lithium	100+	Chile	<1
Tellurium	50–100	Peru	<1
Germanium	5–50	China	<1
Vanadium	100+	China	<1
Tungsten	5–50	China	10–25
Molybdenum	50–100	China	25–50
Selenium	50–100	Chile	<1
	Critical elements of international importance		
Hafnium	5–50	Australia	<1
Nickel	50–100	Australia	>50
Bismuth	5–50	China	<1
Strontium	5–50	China	<1
Barium	/[c]	/[c]	<1
Magnesium	100+	Oceans	25–50
Manganese	5–50	South Africa	>50
Titanium	100+	China	>50
	Critical elements of national importance		
Copper	50–100	Canada	>50
Cadmium	5–50	India	10–25
Silver	5–50	Peru	>50
Tin	5–50	China	>50
Mercury	50–100	Mexico	10–25
Thorium	/[c]	/[c]	/[c]
Arsenic	5–50	China	<1
Yttrium	100+	China	<1
Rubidium	/[c]	/[c]	<1
Caesium	/[c]	/[c]	<1
Zirconium	50–100	Australia	<1
Chromium	100+	Kazakhstan	>50
Rhenium	50–100	Chile	>50
Boron	100+	Turkey	<1
Thallium	50–100	North America	<1

[a]Based on the current known reserves and rate of use [ref. 17–20].
[b]Data adapted from ref. 5,6.
[c]Data not available.

The reserves of many critical elements are confined to a limited number of locations worldwide (Table 5.1). This results in a feedstock vulnerable to geopolitical issues, conflicts and international monopolies. In recent years China has dominated the world's supply of rare earth elements (REE).[22] Diplomatic disputes between China and Japan resulted in limited exports of REE to Japan in 2009.[23] Japan is one of the world's largest users of rare metals, mainly in the production of electronics and consumer goods. The repercussion of restricting the supply of raw materials could have significant and negative economic effects on Japan's economy.[23]

Elements vital to the synthesis of pharmaceuticals also hold similar risks, for example 89% of the world's platinum group metals (PGM) are sourced from South Africa.[9,14,22] Although no restrictions in supply are foreseen in the short term, industrial disputes at PGM mining operations in South Africa (2012) have highlighted the importance of making better use of the elements currently within the supply chain. Australia is also a nation with significant mineral reserves; however, over the next decade they are facing a shortage of skilled labour to work in the mining sector and this is likely to add further pressure or risk to the supply of certain elements. Whilst traditional supply of elements is in decline, their demand and value are rapidly increasing. This is demonstrated in the increase in the production rates and prices of PGM (Figure 5.2).

It is therefore essential that the pharmaceutical industry assesses which elements are required for their continued prosperity and whether these have high risk of restrictions. Life cycle assessments as a tool can enable us to better understand the movement of metals through the product chain and highlight where losses of these valuable resources occur. Suitable green technologies must then be applied to recover these elements.[25]

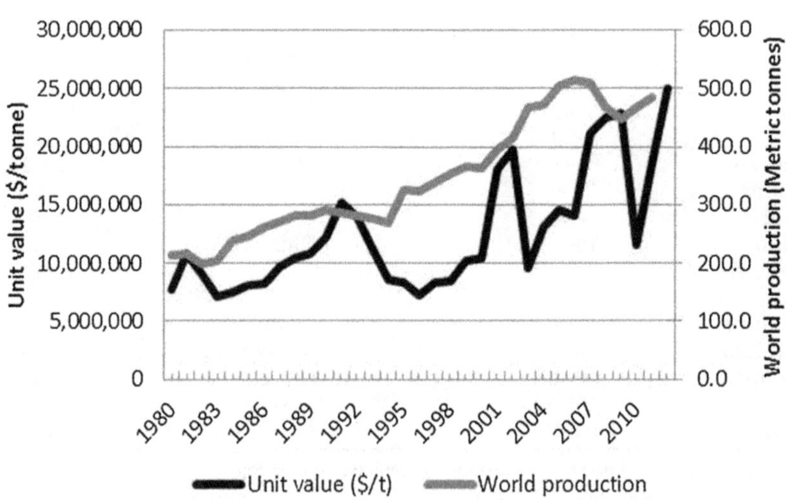

Figure 5.2 Production values and market value of PGM since 1980 (data taken from ref. 24).

5.1.3 Importance to the Pharmaceutical Industry

Many key chemical transformations that are conducted by both medicinal and process chemists within the pharmaceutical industry require elements that have been regarded as critical in Table 5.1 (Figure 5.3). The concept of elemental sustainability is of paramount importance for the long-term prosperity of the pharmaceutical industry. The pharmaceutical industry relies on the use of critical element catalysts to undertake many of the key chemical transformations highlighted in Figure 5.3.[26,27]

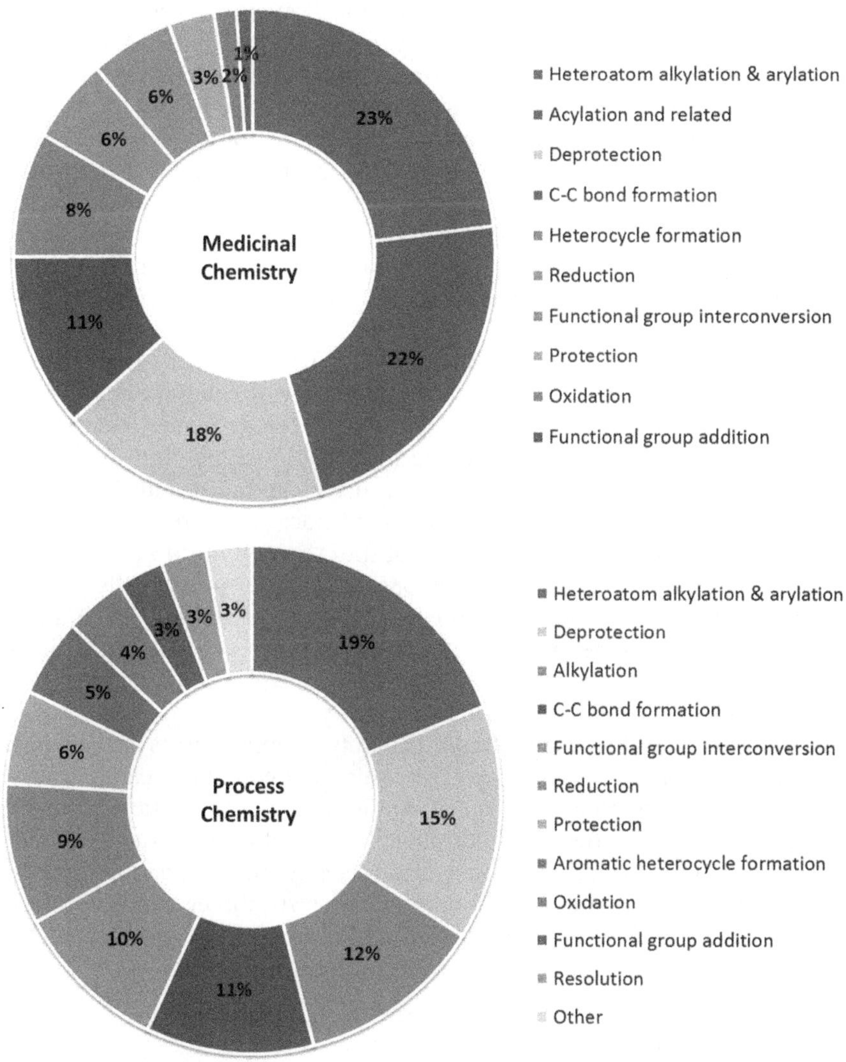

Figure 5.3 Key transformations in medicinal and process chemistry. This figure starts with heteroatom alkylation & arylation in the 12 o'clock position on the chart, and then moves clockwise through the transformations as listed above (data taken from ref. 26, 27).

The two main areas where this industrial sector can have a significant impact are in the use and recovery of these elements within their processes (Figure 5.1). For many elements, sustainable supply chains currently do not exist or are uncertain. A close relationship is required between extractors, manufacturers, users (pharmaceutical sector) and those involved in elemental recovery. Such strategic partnerships could help the pharmaceutical industry move towards a circular economic model. This will require significant investment in research and development, and better dialog across the whole elemental supply chain.

An example of critical element-catalysed reactions is the use of palladium or ruthenium catalysts for C–C bond formation (alkylation, cross-coupling and nucleophilic additions). Such materials are highly efficient but recovery of spent catalyst is of paramount importance in order to move towards a circular economy and close resource loops. Many companies that manufacture catalysts have adopted the strategy of renting or buying these back after use. This ensures a constant supply of feedstock for recovery and regeneration, thus aiding in closing the elemental loop. Other important reactions, including reductions and oxidations, frequently utilize critical elements, such as platinum, palladium, rhodium, nickel, osmium, manganese and cerium.

It is important for the pharmaceutical industry to work with catalyst manufacturers to ensure that the next generation of catalysts is more efficient and utilizes less material. The use of heterogeneous or supported catalysts is seen as one way to aid the recovery of metals, but leaching of elements and therefore the active sites must be reduced and, if possible, eliminated. It will become increasingly important to develop efficient methods for the recovery of elements that find their way into waste systems.

The utilization of metals in a nanoparticulate form may offer further potential for efficient use of elements by the pharmaceutical industry. Nanoparticles are species between 1–100 nm in size, and the properties of such materials differ to bulk metals. Nanoparticles typically have large surface area to volume ratios and high catalytic activity.[28] These properties of nanoparticles are of significant importance to many scientific fields and could be one potential method for reducing critical element content in catalysts.

The challenges of elemental sustainability and the utilization of critical elements are of significant importance to the pharmaceutical industry. Holistic strategies for elemental use and recovery must be adopted, alongside the development of new technologies (catalysts and even new reactions or chemistries) and strategic partnerships to meet these significant challenges that the chemical industry faces in the future.

5.2 Chapter Summary

- Elemental sustainability is the concept whereby all elements within the periodic table are guaranteed for use by both current and future generations.

- Holistic approaches to resource utilization (including new sustainable routes and strategies for recovery and reuse) need to be adopted to meet the significant challenges of elemental sustainability.
- Elements with significant supply risk issues which, if restricted, could harm a nation's economy (or a company's business) are considered to be "critical".
- The application of a closed loop or "cradle-to-cradle" strategies for the production, utilization and recovery of products or resources.
- Greater awareness of the elements and why they are suitable for some reactions will be an essential quality of the future medicinal chemist.

References

1. A. J. Hunt, T. J. Farmer and J. H. Clark, Elemental Sustainability and the Importance of Scarce Element Recovery, in *Element Recovery and Sustainability*, ed. A. J. Hunt, Royal Society of Chemistry, London, 2013, pp. 1–28.
2. M. Z. Jacobson and M. A. Delucchi, *Sci. Am.*, 2009, **11**, 58.
3. House of Commons, Science and Technology Committee, Strategically important metals, Volume I: Report, together with formal minutes, oral and written evidence, 4 May 2011.
4. http://www.bgs.ac.uk/research/highlights/2010/peakMetal.html (Last accessed 11th January 2014).
5. T. E. Graedel, J. Allwood, J.-P. Birat, M. Buchert, C. Hagelüken, B. K. Reck, S. F. Sibley and G. Sonnemann, *J. Ind. Ecol.*, 2011, **15**, 355–366.
6. T. E. Graedel, J. Allwood, J.-P. Birat, B. K. Reck, S. F. Sibley, G. Sonnemann, M. Buchert, C. Hageluken, Recycling Rates of Metals - A Status Report, A Report of the Working Group on the Global Metal Flows of the International Resource Panel, UNEP, 2011.
7. J. Dodson and H. Parker, Mining municipal waste: prospective for elemental recovery, in *Element Recovery and Sustainability*, ed. A. J. Hunt, Royal Society of Chemistry, London, 2013, pp. 220–258.
8. R. L. Moss, E. Tzimas, H. Kara, P. Willis and J. Kooroshy, *Critical Metals in Strategic Energy Technologies*, European Commission, Joint Research Centre Institute for Energy and Transport, 2011.
9. U.S Department of Energy, Critical Element Strategy, 2010.
10. OECD, Critical Metals and Mobile Devices, OECD Environment Directorate, 2010.
11. National Research Council of the National Academies, Minerals, Critical Minerals, and the U.S economy, 2007.
12. United Nations Environment Programme & United Nations University, Critical Metals for Future Sustainable Technologies and their Recycling Potential, 2009.
13. APS Panel on Public Affairs & the Materials Research Society, Energy Critical Elements: Securing Materials for Emerging Technologies, 2011.

14. http://ec.europa.eu/enterprise/policies/raw-materials/files/docs/report-b_en.pdf (Last accessed 11th January 2014).
15. http://www.bgs.ac.uk/mineralsuk/statistics/riskList.html (Last accessed 11th January 2014).
16. http://www.nistep.go.jp/achiev/ftx/eng/stfc/stt027e/qr27pdf/STTqr2704.pdf (Last accessed 11th January 2014).
17. S. E. Kesler, Proceedings for a Workshop on Deposit Modelling, *Mineral Resource Assessment, and Their Role in Sustainable Development*, ed. J. A. Briskey and K. J. Schulz, USGS, USA, 2007, **1**, pp. 55–62.
18. K. Salazar, Mineral commodity summaries 2013: U.S. Geological Survey, U.S. Geological Survey, 2013.
19. T. J. Brown, R. A. Shaw, T. Bide, E. Petavratzi, E. R. Raycraft and A. S. Walters, *World mineral production 2007–11*, British Geological Survey, 2013.
20. C. Rhodes, *Chem. Ind*, 2008, **16**, 21.
21. L. Talens Peiró, G. Villalba Méndez, R. U. Ayres, *Rare and Critical Metals as By-products and the Implications for Future Supply*, 2011, INSEAD.
22. D. Kramer, *Phys. Today*, 2010, **63**, 22.
23. K. Halada, K. Ijima, M. Shimada and N. Katagiri, *J. Jpn. Inst. Metals*, 2009, **73**, 151.
24. T. D. Kelly and G. R. Matos, Historical statistics for mineral and material commodities in the United States (2013 version): U.S. Geological Survey Data Series 140, 2013, http://minerals.usgs.gov/minerals/pubs/historical-statistics/ (Last accessed 11th January 2014).
25. J. R. Dodson, A. J. Hunt, H. L. Parker, Y. Yang and J. H. Clark, *Chem. Eng. Process.*, 2012, **51**, 6926.
26. S. D. Roughley and A. M. Jordan, *J. Med. Chem.*, 2011, **54**, 3451–3479.
27. J. S. Carey, D. Laffan, C. Thompson and M. T. Williams, *Org. Biomol. Chem.*, 2006, **4**, 2337–2347.
28. D. Astruc, F. Lu and J. R. Aranzaes, *Angew. Chem. Int. Ed.*, 2005, **44**, 7852–7872.
29. A. J. Hunt and T. J. Farmer, Elemental sustainability for catalysis, in *sustainable catalysis*, ed. M. North, Royal Society of Chemistry, London, 2015.

CHAPTER 6

Presence, Fate and Risks of Pharmaceuticals in the Environment

KLAUS KÜMMERER

Sustainable Chemistry and Material Resources, Institute of Sustainable and Environmental Chemistry, Faculty of Sustainability, Leuphana University of Lüneburg, Germany
Email: Klaus.Kuemmerer@uni.leuphana.de

6.1 Introduction

The history of the pharmaceutical sciences is a story of never ending creation of new molecules and searching for new mechanisms for the treatment of illness. It has come to light that there may also be unwanted side effects when pharmaceutically active compounds are administered to human beings and animals. The products of the pharmaceutical industry are all around us in everyday life. They increase the standard of living.

The production and manufacturing of pharmaceuticals was associated for a long time with heavy pollution of the environment, as was the case with the production of chemicals.

In general, there has been tremendous progress in the treatment and prevention of waste and effluents associated with the synthesis, production, manufacturing and formulation of pharmaceutical products. Thereby the negative impact on the environment has been reduced by the prevention of emissions into air, water and soil by technical and organizational measures.

RSC Green Chemistry No. 46
Green and Sustainable Medicinal Chemistry: Methods, Tools and Strategies for the 21st Century Pharmaceutical Industry
Edited by Louise Summerton, Helen F. Sneddon, Leonie C. Jones and James H. Clark

Published by the Royal Society of Chemistry, www.rsc.org

However, such approaches are often not yet applied in so-called developing countries.[1]

The presence of pharmaceuticals in the environment is now on the worldwide political agenda as these compounds represent a severe obstacle to sustainable water quality management. In addition, risks to humans cannot be discounted and effects in wildlife have already been observed[2] in lab trials at environmental contamination levels.[3,4] The presence of active compounds in wildlife has been shown.[5] Therefore, the presence of metabolites and transformation products in the environment and the risk resulting from these compounds is of growing interest in this context.

6.2 Active Ingredients, Adjuvants, Metabolites and Transformation Products

Pharmaceutically active ingredients may be present in different forms depending on the pH (Figure 6.1). Even small changes in the chemical structure may have a significant impact on solubility and polarity, *i.e.* extractability, analysis and fate.

Active pharmaceutical ingredients (APIs) as well as adjuvants may reach the environment as active parent compounds because they are often incompletely metabolized. They may also be excreted as metabolites, which might be more or less active. The proportion of the parent compound that remains unmetabolized depends on the parent compound itself and can range from nearly 100% (*e.g.* the anti diabetic drug metformin) to just 5%.[7]

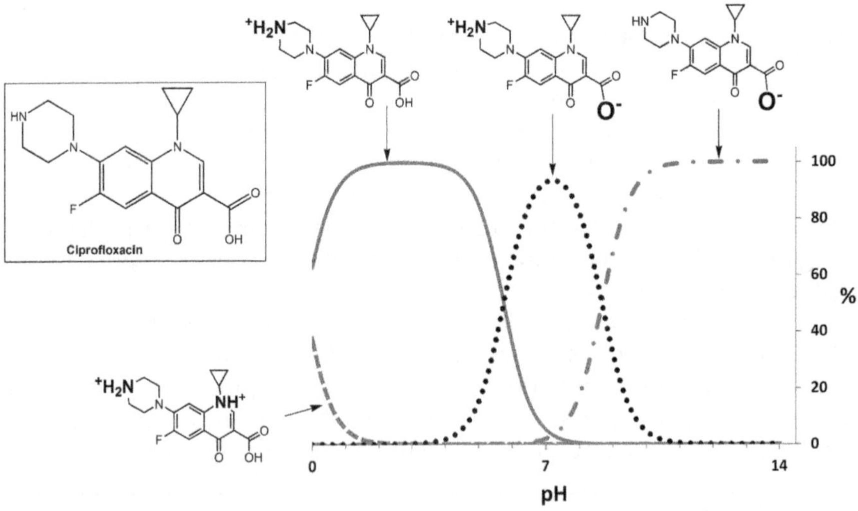

Figure 6.1 Zwitterionic character of ciprofloxacin. Depending on pH, different chemical species are present. (Calculator Plugins were used for structure property prediction and calculation, Marvin 4.1.1, 2006, ChemAxon (http://www.chemaxon.com), reproduced from ref. 6 with kind permission of Springer.)

Some medicines contain molecules based on proteins ("biopharmaceuticals"). The first and best-known example was recombinant human insulin. The environmental relevance of biopharmaceuticals is not yet clear and until now they have not been the focus of environmental research and risk management. However, structurally related compounds such as plasmids have been found in the environment. Furthermore, it is known that the protein structures[1] known as prions are very stable.[8] Most biopharmaceuticals are modified and/or small molecules are attached to them. Therefore, it can be expected that they will behave differently to the pure unmodified proteins. Little is known on their environmental fate and effects.

Often it is assumed that metabolism and transformation of pharmaceuticals reduces the toxicity of the API. In some cases, however, metabolism (*e.g.* in the case of pro-drugs) and transformation by technical treatment processes, such as advanced oxidation, photolysis and others, as well as (bio)chemical reactions in the environment can lead to compounds of higher activity than the parent compound.[9]

Transformation products are frequently poorly characterized and may only be known by their structural formula—if identified at all. Most often these transformation products are not available on the market. Therefore, knowledge of their fate and effects is largely unknown and is unlikely to be fully assessed in the near future. Unexpected effects are possible. This can be illustrated by the case of the fungicide tolylfluanid. Parallels can be drawn between agrochemicals and APIs as APIs are also transformed in the environment and transformation products may reach ground water and potable water. Application of the fungicide tolylfluanid in soil results in a transformation product, *N,N*-dimethylsulfamide. Dimethylsulfamide was found in ground water and surface water in the range of 100–1000 ng L^{-1} and 50–90 ng L^{-1}, respectively.[10] Dimethylsulfamide, owing to its high mobility in soil and water, can enter the drinking water treatment process where it cannot be removed.[9] During ozonation in the treatment plant, about 30–50% of dimethylsulfamide is converted to the carcinogenic nitrosodimethlyamide (NDMA, Figure 6.2).

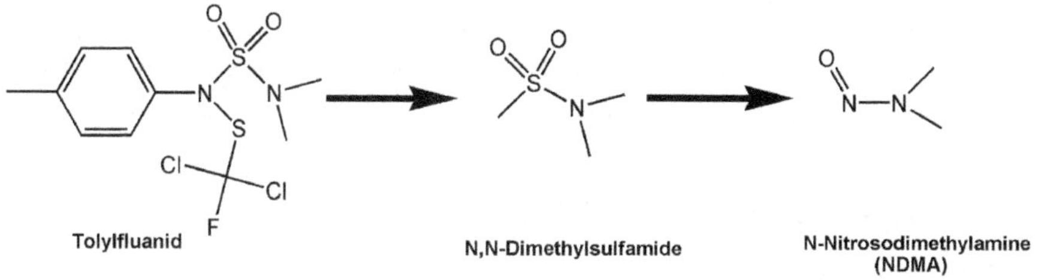

Figure 6.2 Formation of NDMA by ozonation of the transformation product (*N,N*-dimethylsulfamide) of the pesticide tolyfluanid by ozonation in drinking water treatment.

6.3 Introduction into the Environment

There are no data available on the total worldwide use of pharmaceuticals and life-style drugs. The consumption and application of pharmaceuticals may vary considerably from country to country.[12–14] Population growth and improving living standards and access to healthcare as well as increased consumption of meat in the case of veterinary pharmaceuticals lead one to expect that such introduction of pharmaceutically active compounds, adjuvants and contrast agents into the environment will increase.[15]

After excretion, human pharmaceuticals and metabolites enter into wastewater. They are often not removed within wastewater treatment—if it is in place at all. Until recently, there wasn't any knowledge of the fact that pharmaceuticals, contrast media and others constitute a new type of environmental pollution and a possible health risk for the consumer and the environment.

As far as studies are available, it has been revealed that hospitals are, in terms of load, not the main source of input of pharmaceuticals into the environment but instead private households are.[16] This is particularly the case for countries where many pharmaceuticals can be bought without a prescription.

The compounds that are used for veterinary purposes within treatment, prophylaxis or growth-promoting for husbandry and pets are excreted by the animals and will enter the soil if the manure is used as fertilizer or for the amendment of soil,[17] or will reach soils and other surfaces directly if they are administered to pets. From there they can trickle down to ground water or may be washed off from soil by rain reaching surface water.

Furthermore, it has been found that despite good manufacturing practice and corporate sustainable responsibility programs, the production and manufacturing of pharmaceutical products still results in the input of pharmaceuticals and contrast agents into the environment.[18–20] This applies not only for developing countries where concentrations of up to 32 mg L^{-1} of the antibiotic ciprofloxacin (and other active ingredients in nearly the same concentration) were found in drinking water despite the presence of a sewage treatment plant.[1] In Europe as well as in the United States of America it has been found that the concentrations of pharmaceuticals are elevated in receiving waters when a pharmaceutical plant delivers its wastewater to a sewage treatment plant and the sewage treatment plant delivers its effluent to these receiving waters.[18–20] Another minor source is effluents from landfill as date-expired medicines are often disposed of with household waste.[21]

6.4 Presence in the Environment

In the meantime the research on the presence, fate, effects, and risks of pharmaceuticals and contrast agents has triggered more than 15 000 scientific publications. These compounds are known as 'micro-pollutants' because of the lower μg L^{-1} level at which they have been detected.

The detection of pharmaceuticals shows that the parent compound or conjugates can pass sewage treatment plants and reach receiving waters that may be used for drinking water production.[6,20–23]

It is possible that some APIs and adjuvants have been present in the environment for years or even decades. A lack of information about the presence of such micro-pollutants is largely due to a shortage of appropriate analytical methods making compounds difficult to detect. The broader availability of LC-MS and LC-MS/MS in particular permits the detection of polar compounds, such as most pharmaceuticals, metabolites and transformation products, that have not previously been amenable to analysis. Today, there is evidence of the occurrence of some 180 different drugs (parent compounds) and many metabolites in sewage treatment plant effluent, surface water and ground water and the number will increase with the development of new analytical methods.

Numerous studies have shown that a variety of pharmaceuticals are often associated with wastewaters and are present in the vicinity of municipal wastewater discharges. The concentrations of pharmaceuticals in hospital wastewater (in some cases up to a few hundred $\mu g\ L^{-1}$) are higher than in municipal sewage.[11,22–24] However, hospitals are responsible for less than 5% of the total load of pharmaceuticals found in municipal sewage due to the relatively small volume emitted.[16] In most cases, the concentrations of pharmaceuticals in surface waters being detected were in the $ng\ L^{-1}$ range, in rare cases in the low $\mu g\ L^{-1}$ range. They can even be found in the arctic environment.[25]

Some APIs have even been detected in drinking water. APIs have also been detected in sewage sludge, manure and soil. Owing to their persistence, some pharmaceuticals are used by hydrologists as tracers for the assessment of anthropogenic impact on waters.

6.5 Fate

Only a few APIs, such as quinolones and tetracyclines, are adsorbed into sewage sludge or to soil. Thereby they can accumulate in sewage sludge and soil (Figure 6.3). However, there is little knowledge about this.

In general, there is a lack of knowledge about the occurrence, fate and activity of metabolites of pharmaceuticals in the environment. An important question to be addressed is whether or not metabolites such as glucoronides, methylates, glycinates, acetylates, and sulphates are still active. It has been found that some compounds (*e.g.* conjugates of sulfamethoxazole and ethinylestradiol) can be cleaved by bacteria during sewage treatment and in soils releasing the active parent compound.[26] Human metabolites, *e.g.* of ibuprofen and carbamazepine, have not only been detected in sewage treatment plants but also in the receiving rivers and drinking water.[22,23]

After their excretion and release into the environment, both parent compounds and metabolites can undergo further structural changes by a variety

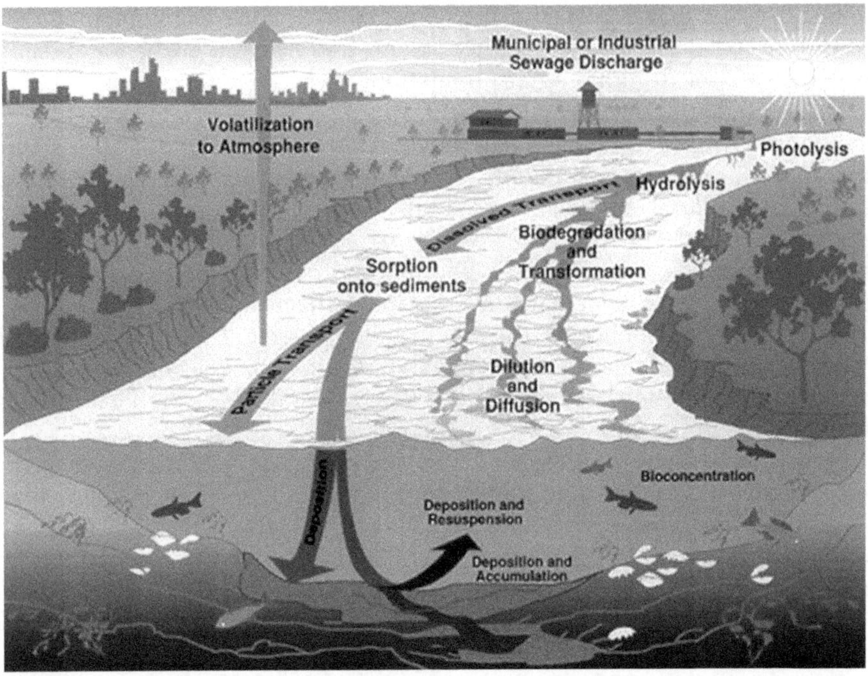

Figure 6.3 The fate of pollutants in the aquatic environment (reproduced from US Geological Survey[40]).

of biotic and non-biotic processes, including photolysis, hydrolysis and biotransformation (Figure 6.3). It has been revealed that not just the parent compounds and metabolites should be of interest but also stable products of incomplete degradation within sewage treatment and the environment,[27–30] and also in the treatment of water in order to produce safe drinking water.[27,31] Most often the formation of transformation products increases the number of chemical entities derived from APIs and adjuvants—a factor of up to five or even tenfold or more per parent compound has been revealed.

6.6 Risks and Risk Assessment

The concentrations of most APIs measured in the aquatic environment are too low to trigger acute effects in humans. One would have to consume several thousand liters of water per day based on the therapeutic dosage. However, some question marks remain; as such, a comparison does not account for long-term ingestion and is based on a single compound approach excluding metabolites and transformation products. It does not include the possible contribution to the increasing number of cases of allergies due to the exposure to multiple chemicals over longer periods. Furthermore, it is known that children and elderly people may be more sensitive. As there are several thousand active ingredients on the market, in addition to the metabolites and transformation products, their analysis is quite challenging, as is risk assessment.

Groups of compounds that need special attention are: (i) antibiotics due to the possible contribution to resistance; (ii) hormones, due to their effects already observed in very low concentration; and (iii) some cytotoxics that can cause cancer. However, for other APIs such as diclofenac and psychoactive drugs effects on wildlife (fish, vultures) may be revealed as well.

Hormones are the main active ingredients in contraceptive pills. In particular, removal of EE2 (17-alpha-ethynylestradiol) from waste water is often nowhere near complete, as has been demonstrated by the findings of Lamoree *et al.*[32] There is evidence that environmentally relevant concentrations of estrogens in surface waters can affect wild fish populations.[33] Recently, an aquatic predicted no-effect concentration (PNEC) of 0.35 ng L^{-1} for EE2 in order to adequately protect organisms in surface waters was recommended by Caldwell *et al.*[34] EE2, 17-beta-estradiol (a naturally occurring hormone) and the pain killer diclofenac were included in the watch list of the EU water framework directive in 2013 in order to be included in routine surface water monitoring.[35]

The European Medicines Agency (EMA, formerly EMEA;[36]) guidelines on the environmental risk assessment of medicinal products for human use states an action limit of 0.01 μg L^{-1}, and the guidelines of the US FDA[37] give a predicted environmental concentration (PEC) of 1.0 μg L^{-1}. However, drugs with endocrine activity are exempted from this limit in the EMA guideline; instead, a tailored environmental risk assessment has to be carried out each time.

Both guidelines are based on a single compound approach for risk assessment, which implies some shortcomings. In real world situations there is always a mixture of several compounds (parent compounds, other chemicals and transformation products thereof). As known from pharmacy, the risk of unwanted side effects increases with the number of active compounds taken at the same time (polypharmacy). There are several concerns about polypharmacy, such as increased adverse drug reactions and drug–drug interactions.[38]

Furthermore, there is no scientific basis to compare effects resulting from a high-dosage short-time ingestion with a life-long uptake, *e.g.* in drinking water. For the latter case, no data is available. Many antineoplastic drugs are used in low quantities. Therefore, their predicted PECs are often below the above-mentioned action limits. However, some antineoplastic compounds that interact directly with DNA possess high activity, even in low environmental concentrations.[39] This fact is not yet being addressed by the EMA and FDA guidelines.

The input of active ingredients and adjuvants of medicines is a major obstacle for a sustainable management of water resources on a global scale. The technical measures taken or under discussion in order to reduce the input of pharmaceutically active compounds and adjuvants have several severe limitations. This is an approach that may make sense in some cases in the short term. However, it has the typical 'end-of-the-pipe' approach issues and is not sustainable. Furthermore, it is, not applicable or affordable on a global scale. Therefore, more prudent use of pharmaceuticals, proper handling of waste (*e.g.* out of date medicines), and a critical attitude to lifestyle drugs are important measures for reduction of the input of pharmaceuticals into the environment. As these methods have limitations, further measures at the beginning of the pipe are urgently needed (see Chapter 7 'Benign by Design').

6.7 Summary

- Active pharmaceutical ingredients (APIs) may reach the environment as active parent compounds because they are most often incompletely metabolized.
- Metabolism (*e.g.* in the case of pro-drugs) and transformation by technical treatment processes, such as advanced oxidation and photolysis, as well as (bio)chemical reactions in the environment can lead to more, often unknown, compounds of possibly even higher activity than the parent compound.
- The environmental relevance of biopharmaceuticals is not yet clear and until now they have not been in the focus of environmental research and risk management.
- The topic now is on the political agenda worldwide as these compounds represent a severe obstacle to sustainable water quality management and as risks to humans cannot be excluded—as for wildlife, effects have already been observed.
- 'End of the pipe' solutions for prevention of input of pharmaceuticals into the environment have severe drawbacks. Therefore, further measures at the beginning of the pipe are urgently needed.

References

1. D. G. Larsson, C. de Pedro and N. Paxeus, *J. Hazard. Mater.*, 2007, **148**, 751.
2. J. L. Oaks, M. Gilbert, M. Z. Virani, R. T. Watson, C. U. Meteyer, B. A. Rideout, H. L. Shivaprasad, S. Ahmed, M. J. Chaudhry, M. Arshad, S. Mahmood, A. Ali and A. A. Khan, *Nature*, 2004, **427**, 630.
3. T. Brodin, J. Fick, M. Jonsson and J. Klaminder, *Science*, 2013, **339**, 814.
4. R. Triebskorn, H. Casper, V. Scheil and J. Schwaiger, *Anal. Bioanal. Chem.*, 2007, **387**, 1405.
5. G. McEneff, L. Barron, B. Kelleher, B. Paull and B. A. Quinn, *Sci. Total Environ.*, 2014, **476–477**, 317–326.

6. *Pharmaceuticals in the Environment: Sources, Fate, Effects and Risks*, ed. K. Kümmerer, 3rd edn, Springer, Berlin Heidelberg, 2008.
7. K. Kümmerer and A. Henninger, *Clin. Microbiol. Infect.*, 2003, **9**, 1203.
8. S. J. Collins, V. A. Lawson and C. L. Masters, *Lancet*, 2004, **363**(9402), 51.
9. A. Magdeburg, D. Stalter, M. Schlüsener, T. Ternes and J. Oehlmann, *Water Res.*, 2014, **50**, 35.
10. C. K. Schmidt and H. J. Brauch, *Environ. Sci. Technol.*, 2008, **42**, 6340.
11. K. Kümmerer, Emerging Contaminants, in *Treatise on Water Science*, ed. P. Wilderer, Elsevier, Amsterdam, 2011, vol. 3, p. 69.
12. United Nations *World Contraceptive Patterns* 2013. http://www.un.org/en/development/desa/population/publications/pdf/family/worldContraceptivePatternsWallChart2013.pdf (access 17 January 2014).
13. H. Goossens, M. Ferech, S. Coenen and P. Stephens, *Clin. Infect. Dis.*, 2007, **44**, 1091.
14. A. K. Sarmah, M. T. Meyer and A. B. A. Boxall, *Chemosphere*, 2006, **65**, 725.
15. *Green and Sustainable Pharmacy*, ed. A. M. Van der, G. Kommer, K. Kümmerer, M. Hempel, Springer Heidelberg Dordrecht, London, New York, 2010, p. 201.
16. A. Schuster, C. Hädrich and K. Kümmerer, *Water, Air, Soil Pollut.: Focus*, 2008, **8**, 457.
17. G. Hamscher, S. Sczesny, H. Höper and H. Nau, *Anal. Chem.*, 2002, **74**, 1509.
18. N. Gilbert, *Nature*, 2011, **476**, 265.
19. USGS 2010: http://toxics.usgs.gov/highlights/PMFs.html (access 9 July 2015), courtesy of the US Geological Service.
20. J. P. Phillips, S. G. Smith, D. W. Kolpin, S. D. Zaugg, H. T. Buxton, T. E. Furlong, K. Esposito and B. Stinson, *Environ. Sci. Technol.*, 2010, **44**, 4910.
21. J. V. Holm, K. Ruegge, P. L. Bjerg and T. H. Christensen, *Environ. Sci. Technol.*, 1995, **29**, 1415.
22. T. Heberer, *Toxicol. Lett.*, 2002, **131**, 5.
23. K. Kümmerer, *Annu. Rev. Environ. Resour.*, 2010, **35**, 57.
24. European Environmental Agency (EEA), *EEA Technical report No 1*, Copenhagen, 2010.
25. R. Kallenborn, J. Fick, R. Lindberg, K. M. Nielsen, M. Tysklind, T. Vasskog, in *Pharmaceuticals in the Environment: Sources, Fate, Effects and Risks*, ed. Kümmerer, 3rd edn, Springer, Berlin Heidelberg, 2008, p. 61.
26. K. Schauss, A. Focks, H. Heuer, A. Kotzerke, H. Schmitt, S. Thiele-Bruhn, K. Smalla, B.-M. Wilke, M. Matthies, W. Amelung, J. Klasmeier and M. Schloter, *TrAC, Trends Anal. Chem.*, 2009, **28**, 612.
27. S. Zühlke, U. Dünnbier and T. Heberer, *J. Chromatogr.*, 2004, **1050**, 201.
28. A. Haiß and K. Kümmerer, *Chemosphere*, 2006, **62**, 294.
29. J. Gröning, C. Held, C. Garten, U. Claussnitzer, S. R. Kaschabek and M. Schlömann, *Chemosphere*, 2007, **69**, 509.

30. W. M. M. Mahmoud, A. P. Toolaram, J. Menz, C. Leder, M. Schneider and K. Kümmerer, *Water Res.*, 2014, **49**, 11.
31. L. Tootchi, R. Seth, S. Tabe and P. Yang, *Water Sci. Technol.*, 2013, **13**, 1576.
32. M. H. Lamoree, J. G. M. Derksen, S. C. van der Linden, C. A. Uijterlinde and P. de Voogt, in *Xenobiotics in the Urban Water Cycle*, ed. D. Kassinos, K. Bester and K. Kümmerer, Springer, New York, 2009, **16**, 261.
33. K. A. Kidd, P. J. Blanchfield, K. H. Mills, V. P. Palace, R. E. Evans, J. M. Lazorchak and R. W. Flick, *Proc. Natl. Acad. Sci.*, 2007, **104**, 8897.
34. D. J. Caldwell, R. Mastrocco, T. H. Hutchinson, R. Länge, D. Heijerick, C. Janssen, P. D. Anderson and J. P. Sumpter, *Environ. Sci. Technol.*, 2008, **42**, 7046.
35. EU, Directive 2013/39/EU of the European Parliament and of the Council of 12 August 2013 amending Directives 2000/60/EC and 2008/105/EC as regards priority substances in the field of water policy, *Off. J. Eur. Union*, 2013, **L 226/1** (access 22 December 2013).
36. European Medicines Agency (EMA) (2006) Doc. Ref. EMEA/CHMP/SWP/4447/00. EMEA, London. http://www.ema.europa.eu/docs/en_GB/docment_library/Scientific_guideline/2009/10/WC500003978.pdf.
37. FDA *Guidance for Industry Environmental Assessment of Human Drug and Biologics Applications*. US Department of Health and Human Services. Food and Drug Administration Center for Drug Evaluation and Research (CDER) Center for Biologics Evaluation and Research (CBER) July 1998 CMC 6 Revision 1, http://www.fda.gov/downloads/Drugs/GuidanceComplianceRegulatoryInformation/Guidances/ucm070561.pdf (access 30 July 2013).
38. E. R. Hajjar, A. C. Cafiero and J. T. Hanlon, *Am. J. Geriatr. Pharmacother.*, 2007, **5**, 345.
39. K. Kümmerer, A. Haiß, A. Schuster, A. Hein and I. Ebert, *Environ. Sci. Pollut. Res.*, in press, DOI: 10.1007/s11356-014-3902-8.
40. http://toxics.usgs.gov/regional/emc/transport_fate.html (access July 2015), courtesy of the US Geological Service.

CHAPTER 7

Benign by Design

KLAUS KÜMMERER

Sustainable Chemistry and Material Resources, Institute of Sustainable and Environmental Chemistry, Faculty of Sustainability, Leuphana University of Lüneburg, Germany
Email: Klaus.Kuemmerer@uni.leuphana.de

7.1 Introduction

In general, pharmaceuticals are not fully metabolized or mineralized in the body of the target organism after their administration. In fact, parent compounds and metabolites are excreted and human excretions reach sewage systems. These parent compounds and metabolites are often not fully eliminated during sewage treatment and therefore enter the aquatic environment, and may eventually reach drinking water. Excretions of animals containing active pharmaceutical ingredients (APIs), adjuvants and their metabolites may reach the environment directly with the manure if the manure is applied as a fertilizer. The (complete) mineralization of these compounds is generally also low in biogas production plants. In this case, the compounds will reach the soil when the remainder of the biogas production is applied as fertilizer or for land amendment. Furthermore, pets on medication may excrete the APIs into the environment directly and APIs applied in aquaculture are directly introduced into the water cycle.[1,2]

This is demonstrated by the fact that pharmaceuticals are detectable in the environment,[3] (see also Chapter 6, Presence, Fate and Risks of Pharmaceuticals in the Environment). It has been shown that even advanced effluent treatment is not able to fully remove the pharmaceutically active

RSC Green Chemistry No. 46
Green and Sustainable Medicinal Chemistry: Methods, Tools and Strategies for the 21st Century Pharmaceutical Industry
Edited by Louise Summerton, Helen F. Sneddon, Leonie C. Jones and James H. Clark

Published by the Royal Society of Chemistry, www.rsc.org

compounds (APIs) and metabolites—if sewage treatment is in place at all.[4–7] Furthermore, most advanced treatment technologies that are now in use produce stable transformation products that can potentially be even more toxic than the parent compound.[8,9] Risk assessment of these transformation products is complex—if possible at all—as data on their fate and effects in the environment is often unavailable. The transformation products must be detected and identified, which may be further complicated by the fact that they are often not commercially available for testing. Furthermore, such treatments are 'end of the pipe' approaches that are often not sustainable.

7.2 Inherently Safe

Low toxicity in humans is an essential prerequisite for the registration of APIs and adjuvants. Pharmaceuticals that fulfil the needs of their application, (*i.e.* are generally non-toxic and have only low or, preferably, no unwanted side effects) could be considered to be 'inherently safe'. This is one of the core concepts of the development and application of pharmaceuticals. The extension of the concept of inherently safe APIs to sustainability results in a new requirement that chemicals should be completely and readily degraded to harmless products, *i.e.* that they are fully mineralized within a reasonable time on entering the environment. Reducing or designing out unwanted side effects shouldn't just be an issue related to patient safety but also to the environment. In other words, 'eco-pharmacovigilance'[10] also includes the consideration of unwanted effects in the environment and efforts to avoid them. A statistical correlation between the persistence of chemicals and the prevalence of chemicals present in the environment that can initiate cancer has recently been reported.[11] APIs, adjuvants, metabolites and their transformation products can often not be removed from the water cycle once they have entered it. This is for thermodynamic reasons as well as for technical and economical limitations. Therefore, the presence of such compounds in the aquatic environment is seen as one of the biggest obstacles to sustainable water in the future, especially in arid countries where water reuse is of high importance.

7.3 'The End of the Pipe's End'

According to the principles of green and sustainable pharmacy, the functionality of an API should not only include the properties necessary for its application, but also rapid and complete mineralization when it enters the environment. This concept is known as 'Benign by Design', which is a key building block of sustainable pharmacy.[12–14]

However, it goes without saying that we expect an API to be stable, in other words it should be, at least to a certain degree, persistent. Persistence is one of the most important criteria in the environmental assessment of chemicals. A long half-life of APIs, adjuvants, their metabolites and transformation products in the environment results in their wide distribution.

The bigger the time-scales involved, the more uncertainty increases and the more difficult it becomes to test possible effects in lab trials.[15] It is therefore impossible to carry out a reliable risk assessment for such molecules.

7.4 Stability—a Question of Conditions

Readily and fully mineralizable APIs and adjuvants avoid all these problems. However, conventional wisdom assumes that an API needs to be stable to be successful in the market. A closer look reveals that this is not necessarily the case. On the one hand, there are lots of examples demonstrating that good performance of an API is not necessarily in contradiction with good biodegradability in the environment as some readily biodegradable APIs demonstrate (Table 7.1).[12,16–19] An analysis of data from our own biodegradability database shows that 29% of pharmaceuticals from different chemical and therapeutic classes are mineralized in the environment. For example, the anti-epileptic drug valproic acid has been used in high amounts (*e.g.* 69 metric tons in Germany in 2006) for more than 40 years now. It is fully mineralized by environmental bacteria. On the other hand, some well-known "unstable" APIs have been very successful on the market for a long time. Some β-lactam antibiotics, such as amoxicillin, are unstable in solution and so are delivered to the patient as a dry powder that has to be mixed with tap water before administration. After mixing they are stored in the fridge until use. After excretion into wastewater non-biotic and/or biotic processes hydrolyze the intact portion of these APIs. Some penicillins, such as ampicillin, are fully mineralized in sewage treatment plants (STPs), and others, such as piperacillin, are at least deactivated.[20] Other APIs are sensitive to light. They achieve sufficient shelf life by storage in brown glass bottles to prevent photodecomposition. These data demonstrate that mineralization of APIs in the environment and within application is not necessarily a contradiction. From another point of view, stability and degradability are a question of kinetics and thermodynamics, *i.e.* the relation of the

Table 7.1 Biodegradability of selected pharmaceuticals from various classes in different OECD tests. Criterion for ready biodegradability in 301 tests and inherent biodegradability in 302 tests is >60%. Reproduced from ref. 21 with kind permission by Springer.

Active compound	OECD 301D	OECD 301F	OECD 302B
Isosobiddinitrate	>90		
Mesalazin	>90	>90	
Acetylsalicylic Acid		81	
Penicillin V	27		>90
Glufosfamide	53		72
Piracetam		>90	>90
Hydroxamic Acid	50	90	
Valproic Acid	72	78	
Cytarabin	40		>90

Table 7.2 Some selected physical, chemical and microbiological (reaction) conditions along the life cycle of APIs.

Location	Moisture	pH	Temperature	Bacterial density and diversity	Light
Pharmacy	Dry	According to product stability, most often 7	20–25 °C	Very low or not at all (sterile)	Possible, but no direct sunlight
Human body	Wet	Blood: 7.4 Stomach:1.5 (empty stomach, up to 6 after food intake)	36.5 °C	Gut: high, anaerobic	No
Waste water/ sewage treatment plant	Wet	>8	<15 °C	Very high aerobic	No
Surface water	Wet	7	<25 °C	Low	Yes, intensive sun light and UV

energies and the time scales of the different reaction pathways that may compete under certain conditions.

Whilst molecular structure and desired properties will clearly vary greatly from drug to drug, it is fair to say that the paradigm that APIs have to be stable under most conditions as an indispensable property for their application is not true.

The reactivity of a molecule depends on both the properties of the molecule and its environment. Every chemist and pharmacist knows that a molecule may react faster, slower or in a different way when conditions such as pH, redox potential, moisture, temperature, exposure to light, its wavelength and intensity, and presence of other molecules are changed. This is the case along the life cycle of an API including bacterial diversity and density (see Table 7.2).

APIs have to interact within their range of application. By their very nature, they may interact with the biological environment; however, they must remain stable in order to interact with their target. The challenge is to design a molecule in such a way that its lifetime is sufficient for its use but short enough under environmental conditions.

7.5 Structure Matters—Benign by Design

The targeted design of an API in order to fulfil the requirements of both application and the environment is a challenging task. Even small changes in the structure of a chemical may have tremendous effects on its properties. Benzene for example, is (after metabolic activation) a carcinogen.[23] It is only slowly biodegradable. The simple insertion of an oxygen atom into a carbon–hydrogen bond results in a new compound, phenol, with very different properties which impacts on its fate in the environment. In contrast to benzene, phenol is not classifiable as to its carcinogenicity to humans (group 3),[23] readily aerobically biodegradable and active against bacteria. Therefore, it has long been used as a disinfectant. It may also be the case for some APIs that compounds of similar structure, with the same mode of action, have different biodegradability (Figure 7.1). The modified structure of glufosfamide compared to ifosfamide (Figure 7.1) resulted in the desired improvement of pharmaceutical properties, *i.e.* the reduction of unwanted side effects. At the same time—without specific intention—the biodegradability was improved tremendously. It is not only attributable to the sugar moiety, but the whole molecule is subsequently more readily biodegradable in the environment. Furthermore, in the case of cytarabine, the impact of the density and diversity of the bacteria being present can be seen (Figure 7.1). In a CBT (closed bottle test, representing surface water) there are a maximum of 1000 colony forming units per millilitre present. In the ZWT (Zahn-Wellens test, representing sewage sludge) the bacterial density is a few grams per litre. The biodegradability of gemcitabine is lower than cytarabine because of the presence of the two fluorine atoms in the sugar moiety. It is well known that organic carbon–fluorine bonds are resistant to biodegradation.[24–26] Both examples (Figure 7.1) have in common that the purpose of the structural modification of the older molecule (5-FU and ifosfamide, respectively) was to improve the pharmaceutical properties and to reduce unwanted side effects in patients. The environmental impact was not necessarily taken into consideration. However, both examples demonstrate clearly that biodegradable APIs can be successful on the market as pharmaceuticals, meeting the requirements to be classified as inherently biodegradable.

As with the simple example of benzene, there are many ways to modify an API's structure. Well-established test systems,[22] empirical data, rules of thumb[24] (Figure 7.2) and computer-based systems for the assessment of the properties of molecules,[25,27] such as biodegradability,[26] are available. Their careful combination allows for the theoretical design of molecules and the assessment of their properties without the need for them to be synthesized. Computer-based systems that assist human experts with the handling of complex knowledge are widely in use within pharmaceutical industries for the optimization of APIs by *in silico* screening.[27] This allows for a very early and comparatively inexpensive evaluation of the environmental properties of candidates. At the same time, undesirable properties such as mutagenicity can be reduced. This can and should also be done with respect to full mineralization of APIs when they reach the environment.

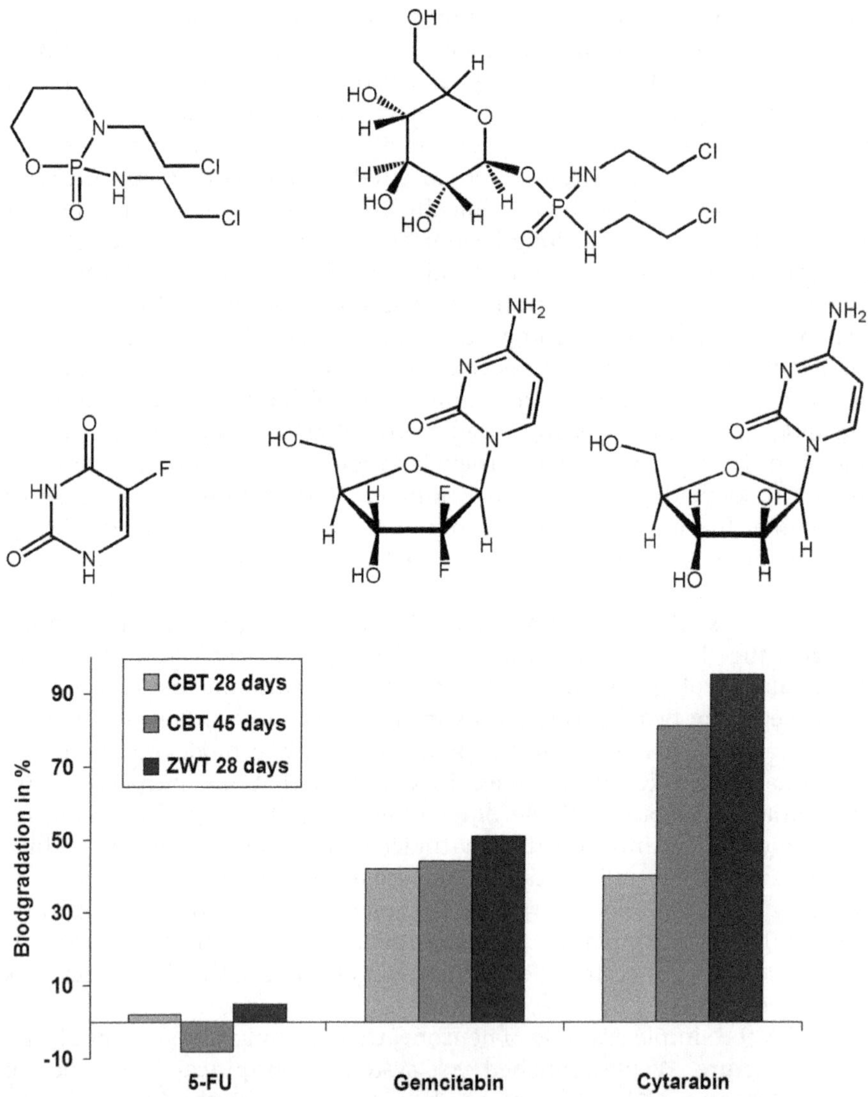

Figure 7.1 Top: ifosfamide (left) and β-D-Glc-IPM (INN: glufosfamid; right). Bottom: biodegradability of structurally related cytotoxics (top from left to right: 5-fluorouracil (5-FU), gemcitabine and cytarabine) in different OECD tests[22] for the assessment of biodegradability (CBT: closed bottle test, OECD 301D; ZWT: Zahn-Wellens test, OECD 302B).
Reproduced from ref. 7 with permission. © Wiley (1997).

The concept of including a goal to develop molecules that are readily and fully mineralized when they reach the environment, *e.g.* the water resource or soil, into their targeted design is called 'Benign by Design'.[12] This approach may result in new chemical entities, thereby creating new economic

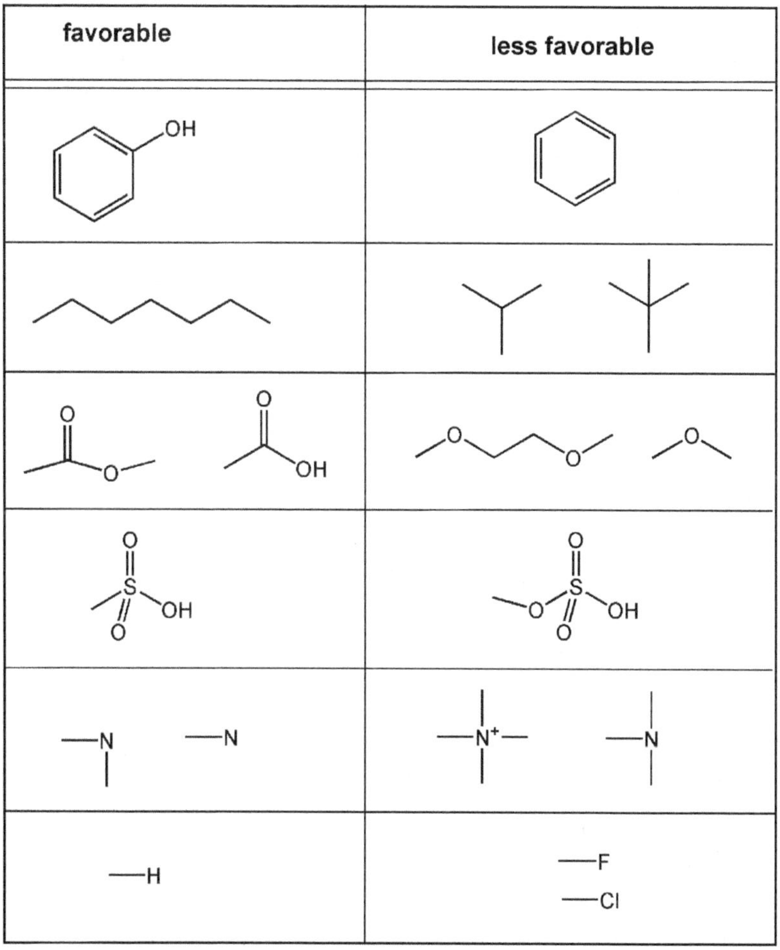

Figure 7.2 Examples of chemical functionalities and their impact on biodegradability.
Reproduced from ref. 21 with kind permission by Springer.

opportunities. Such APIs allow for sustainable use, even in countries where there is only little or no effluent treatment in place. However, some additional knowledge and further experience needs to be gained on the opportunities and limitations of this approach.

It is obvious that few drugs could have such simple modifications made, as shown in Figure 7.2, without losing other vital properties. These structures represent only the very beginning of potentially valuable additions to the list of considerations medicinal chemists have to make. There is also a need to better understand the results of different combinations of such structural elements of new (and old modified) molecules. Furthermore, this new concept also holds the promise of new APIs and products. First steps to put this into practice have already been made.[28–30]

7.6 Summary

- APIs may degrade incompletely in the environment, thus resulting in numerous transformation products. Both parent compounds and transformation products are present in the environment.
- New approaches are necessary to prevent the input of pharmaceuticals into the environment.
- Developing new API molecules and adjuvants that are readily and fully mineralized once they leave the body is such an approach ("Benign by Design").
- This new concept is a challenge but also leads to new APIs and products.

References

1. A. Knobloch, S. A. Mohring, N. Eberle, I. Nolte, G. Hamscher and D. Simon, *J. Vet. Intern. Med.*, 2010, **24**, 384.
2. *Pharmaceuticals in the Environment. Sources. Fate, Effects and Riskm*, ed., K. Kümmerer, Berlin Heidelberg, Springer, 3rd edn, 2008.
3. K. Kümmerer, *Annu. Rev. Environm. Resour.*, 2010, **35**, 57.
4. M. Carballa, F. Omil, J. M. Lema, M. Llompart, C. García and I. Rodriguez, *Water Sci. Technol.*, 2005, **52**, 9.
5. M. J. Capdeville and H. Budzinski, *TrAC, Trends Anal. Chem.*, 2011, **30**, 586.
6. C. M. Coetsier, S. Spinelli, L. Lin, B. Roig and E. Touraud, *Environ. Int.*, 2009, **35**, 787.
7. L. Tootchi, R. Seth, S. Tabe and P. Yang, *Water Sci. Technol.: Water Supply*, 2013, **13**, 1576.
8. E. llés, E. Szabó, Takács, L. Wojnárovits, A. Dombi and K. Gajda-Schrantz, *Sci. Total Environ.*, 2014, **472**, 178.
9. A. Magdeburg, D. Stalter, M. Schlüsener, T. Ternes and J. Oehlmann, *Water Res.*, 2014, **50**, 35.
10. K. Kümmerer and G. A. Velo, *Drug Saf.*, 2006, **29**, 371.
11. N. Pollack, A. R. Cunningham and H. S. Rosenkranz, *Mutat. Res., Fundam. Mol. Mech. Mutagen.*, 2003, **528**, 81.
12. K. Kümmerer, *Green Chem.*, 2007, **9**, 899.
13. *Pharmaceuticals in the Environment. Sources. Fate, Effects and Risk*, ed. K. Kümmerer, Springer Publisher, Berlin Heidelberg New York, 3rd edn, 2008, pp. 433–440.
14. *Green and Sustainable Pharmacy*, ed. K. Kümmerer and M. Hempel, Springer, Berlin, Dordrecht, London, New York, 2010.
15. J. Cairns Jr. and D. I. Mount, *Environ. Sci. Technol.*, 1990, **24**, 154.
16. K. Kümmerer and A. Al-Ahmad, *Acta Hydrochim. Hydrobiol.*, 1997, **25**, 166.

17. K. Kümmerer, A. Al-Ahmad, B. Bertram and M. Wießler, *Chemosphere*, 2000, **40**, 767.
18. A. Thiem, N. Schmift, M. Steiber, F. Sacher, L. Wolf and H. Hoetzl, *Water Resour. Manage.*, 2011, **25**, 1195.
19. P. Guerra, M. Kim, A. Shah, M. Alaee and S. A. Smyth, *Sci. Total Environ.*, 2014, **473**, 235.
20. A. Längin, R. Alexy, A. König and K. Kümmerer, *Chemosphere*, 2009, **75**, 347.
21. K. Kümmerer, in *Green and Sustainable Pharmacy*, ed. K. Kümmerer and M. Hempel, Springer, Berlin, Dordrecht, London, New York, 2010, p. 135.
22. OECD, *Guidelines for Testing of Chemicals, Section 3, Degradation and accumulation*. OECD Publishing. DOI: 10.1787/9789264030213-en; http://www.oecd-ilibrary.org/environment/oecd-guidelines-for-the-testing-of-chemicals-section-3-degradation-and-accumulation_2074577x.
23. http://monographs.iarc.fr/ENG/Classification/ClassificationsAlphaOrder.pdf.
24. R. S. Boethling, E. Sommer and D. DiFiore, *Chem. Rev.*, 2007, **107**, 2207.
25. R. S. Boethling, D. Mackay, *Handbook of Property Estimation Methods for Chemicals – Environmental and Health Sciences*, Lewis Publisher, Boca Raton, 2000.
26. C. Rücker and K. Kümmerer, *Green Chem.*, 2012, **14**, 875.
27. *Computational Toxicology*, ed. S. Ekins, John Wiley and Sons, Hoboken, 2007.
28. T. Rastogi, C. Leder and K. Kümmerer, *Chemosphere*, 2014, **111**, 493.
29. T. Rastogi, C. Leder and K. Kümmerer, *RSC Adv.*, 2015, **5**, 27.
30. T. Rastogi, C. Leder and K. Kümmerer, *Environ. Sci. Technol.*, 2015, **49**, 11756.

CHAPTER 8

From Discovery to Manufacturing: Some Sustainability Challenges Presented by the Requirements of Medicine Development

JOHN HAYLER

GlaxoSmithKline Medicines Research Centre, Gunnels Wood Road, Stevenage, Hertfordshire SG1 2NY, UK
Email: John.Hayler@gsk.com

8.1 Introduction

The discovery and development of a new medicine is a long and complicated endeavour requiring the contribution of multiple teams of people over many years. Medicinal chemists are key members of a discovery team, involved from the conception of an idea through to, usually, selection of a potential drug candidate. At a particular time point, which varies depending on company approach, responsibility for the development of the drug candidate will pass to a development team. The aim of the development process is to demonstrate to the satisfaction of the regulatory authorities that the proposed new medicine is safe, efficacious and of assured quality. Upon completion of the clinical and safety studies, the data is submitted for review in

RSC Green Chemistry No. 46
Green and Sustainable Medicinal Chemistry: Methods, Tools and Strategies for the 21st Century Pharmaceutical Industry
Edited by Louise Summerton, Helen F. Sneddon, Leonie C. Jones and James H. Clark

Published by the Royal Society of Chemistry, www.rsc.org

the form of a Marketing Authorisation Application (MAA, EU), (Japanese) New Drug Application (JNDA, Japan; NDA, USA) or equivalent. If the application is successful the licensing authority will grant approval to market the medicine in their territory.

This chapter will:

- Summarize the aims of the different phases of drug development
- Discuss the ideal characteristics of a manufacturing route to an API
- Discuss the quality requirements for the manufacture of an API
- Consider the potential sustainability impact of these requirements

The emphasis of this chapter is on the chemical development of small molecule active pharmaceutical ingredients (API); that is, a single chemical entity, natural product, or combination of chemicals, which forms the active constituent of a formulated drug product. Biopharmaceuticals, proteins and oligosaccharides are not discussed, although many of the same principles will apply.

8.2 Drug Development

The safety and efficacy of a new medicine is demonstrated through a series of controlled clinical trials, which are summarised in Table 8.1.

Table 8.1 Drug development activities.

Phase	Activities and purpose	Approximate time and cost[a]	API required
Pre-clinical	Enabling non-clinical safety testing (ICH M3) and formulation development to support Phase I		100s g–few kg
Phase I	Human safety data Additional non-clinical safety studies to enable long-term dosing	21.6 months $15.2 m	
Phase IIa	Determine proof of concept Human safety Additional non-clinical safety studies to enable long-term dosing	25.7 months	10s kg
Phase IIb	Dose range finding studies Human safety Additional non-clinical safety studies to enable phase III and registration	$23.4 m	
Phase III	Determination of efficacy Human safety and side effects Non-clinical safety studies to support registration	30.5 months $86.5 m	100s kg–tonnes

[a]Approximate time and cost taken from Laufer *et al.*[1]

A regulatory submission will typically take the form of the Common Technical Document (CTD),[2] which was developed by the International Conference on Harmonisation of Technical Requirements for Registration of Pharmaceuticals for Human Use (ICH).[3] Established in 1990, the ICH is a collaboration between the regulatory authorities and pharmaceutical industries of Europe, Japan and the USA; their mission is "*to achieve greater harmonisation to ensure that safe, effective, and high quality medicines are developed and registered in the most resource-efficient manner.*"

The CTD provides a consistent format for the regulatory submission and agreed standards for activities supporting the different sections. The consistent format has enabled companies to submit to different regulatory authorities in an efficient manner, provided a quicker review process and, *via* the eCTD, facilitated electronic submission, replacing 100 000s of pages of print.[4] The ICH has also agreed guidelines on the activities required to support a regulatory submission, these cover Quality (Q), Safety (S), Efficacy (E) and Multidisciplinary (M) activities and standards. The relevant guidance is referenced through this document, *e.g.* ICH M3 = non-clinical safety studies required to support clinical trials and marketing authorization.[5]

The CTD comprises a series of modules:[2]

- Module 1, regional administrative information (not part of the CTD)
- Module 2, quality overall summary (general introduction, quality summary, non-clinical overview and summary, clinical overview and summary)
- Module 3, quality: M4Q(R1)
- Module 4, safety (non-clinical study reports) M4S(R2)
- Module 5, efficacy (clinical study reports) M4E(R1)

Modules 2.3 (overall summary) and 3 define the quality assurance of the medicine and are commonly referred to as the Chemistry Manufacturing and Controls (CMC) prepared by the chemistry and pharmaceutical development teams; their activities run alongside the development programme.

8.3 Development of the API Commercial Route

8.3.1 Discovery

A synthetic route comprises a series of chemical transformations from defined starting materials to the final molecule. The nature and requirements of a synthetic route change through discovery and development. In the discovery phase, a route should be flexible to enable the preparation of milligram quantities of a diverse set of compounds, although there is a risk that this approach can lead to missed opportunities if the products that are easily obtained are the only ones assessed.[6] An example of a discovery route for the preparation of a series of indole acetic acids for the study of

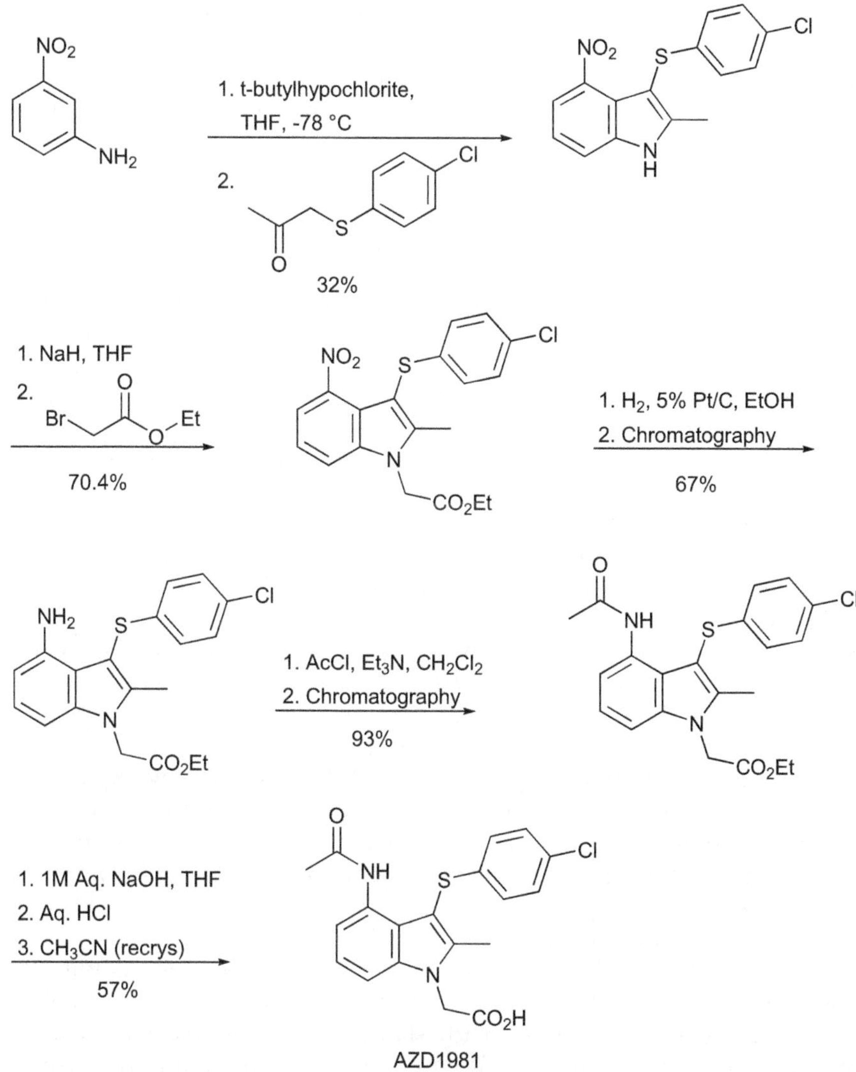

Scheme 8.1 The discovery route to AZD1981.[7]

respiratory inflammation is illustrated in Scheme 8.1, showing the lead compound AZD1981.[7,8]

The route allows for the introduction of diversity (position and structure):

- On the indole, through selection of variously substituted anilines (including halogen for subsequent cross-coupling reactions)
- In the C-3 arylthio ether through variously substituted thiophenols
- At C-2 through the choice of ketone
- On the amino group, *e.g.* amides, sulfonamides *etc.*

37 examples are described in the patent, prepared in amounts up to 250 mg.[7] The efficiency of the route is not a priority and the synthesis shown in Scheme 8.1 has an overall yield of 8%, with a projected process mass intensity (PMI)[9] of 718 kg kg^{-1} API, excluding some parts of the work-up and chromatography.

8.3.2 Ensuring Delivery to Proof of Concept

Once a candidate has been selected, the chemistry team have the advantage of a single target compound to prepare. Their aims are to design and develop the optimum route to the API whilst ensuring its supply to support the development programme. Sufficient API is required for safety assessment studies, formulation development to deliver the drug product and manufacture of clinical trials supplies. The amount of API required to support the different clinical phases for an oral drug product is summarised in Table 8.1 (assuming a dose in the range of mg – ≤2 g). Initial requirements to support development are usually limited to a few kilograms and the discovery route will often be used by first intent if starting materials can be sourced in sufficient quantity.

Some process development will be carried out to improve processing and ensure safe scale-up of the chemistry by replacing unsafe reagent/solvent combinations and addressing potential thermal instabilities and gas evolution. Often the effort is placed on the work-up and isolation of intermediates, *e.g.* developing appropriate quench and washing regimes to remove impurities and developing conditions for product crystallization in place of a work-up comprising a quench, extraction, wash, evaporation and chromatographic purification. Scale-up of chromatography and, in particular, preparative HPLC separation of isomers can effectively allow rapid scale-up for initial kilogram manufacture.[10]

In some cases, an alternative route may be required, for example the supply route to AZD1981 illustrated in Scheme 8.2. Here the core indole is prepared *via* a Makosza indole synthesis of **1**, avoiding cryogenic reaction conditions in stage 1 of the discovery route, eliminating the use of dichloromethane in stage 4 and removing all chromatographic purification. This resulted in at least a doubling of the overall yield.[11]

8.3.3 From Proof of Concept to Manufacturing

With an established supply route in place, the chemistry team can focus on identification of the preferred commercial route of manufacture. A synthetic route should ideally meet multiple criteria and there have been a number of publications assessing these.[12] One approach is the SELECT criteria,[13] which are aligned with many of the 12 principles of green chemistry,[14] as summarised in Table 8.2.

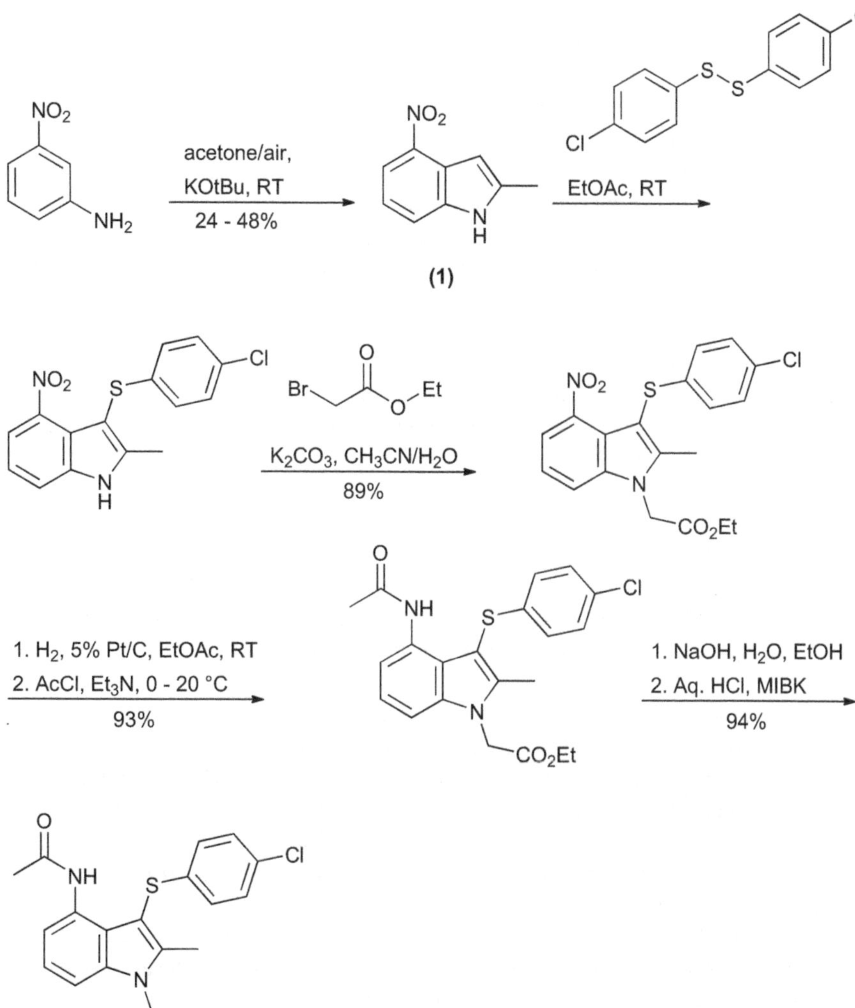

Scheme 8.2 The supply route to AZD1981: 18.7–37% overall yield.[11]

8.3.4 The 12 Principles of Green Chemistry[14]

1. It is better to prevent waste than to clean up waste after it is formed.
2. Synthetic methods should be designed to maximise the incorporation of all materials used in the process into the final product (atom economy).
3. Wherever practicable, synthetic methodologies should be designed to use and generate substances that possess little or no toxicity to human health and the environment.
4. Chemical products should be designed to preserve efficacy of function while reducing toxicity.

5. The use of auxiliary substances (*e.g.* solvents, separating agents) should be made unnecessary wherever possible and innocuous when used.
6. Energy requirements should be recognized for their environmental and economic impacts and should be minimized.
7. A raw material or feedstock should be renewable rather than depleting, wherever technically and economically practicable.
8. Unnecessary derivatization (blocking group, protection/deprotection, temporary modification of physical/chemical processes) should be avoided whenever possible.
9. Catalytic reagents (as selective as possible) are superior to stoichiometric reagents.
10. Chemical products should be designed so that at the end of their function they do not persist in the environment and break down into innocuous degradation products.
11. Analytical methodologies need to be further developed to allow for real-time, in-process monitoring, and control prior to the formation of hazardous substances.
12. Substances and the form of a substance used in a chemical process should be chosen so as to minimize the potential for chemical accidents, including releases, explosions and fires.

As can be seen in Table 8.2, a route meeting the SELECT criteria could fulfil all of the 12 principles of green chemistry and *vice versa*. For example, application of green principles to the design of the synthetic route can help contribute to the second part of the legal criteria, where the use of minimal impact materials (*e.g.* low volatility solvents) will help meet emissions targets set by local environmental legislation, *e.g.* the European Union industrial emissions directive.[15]

Safety considerations were a driver to change the supply route in the synthesis of AZD1981 where the Makosza reaction was conducted above the flash point of acetone in the presence of air and the product nitro-indole (**1**, Scheme 8.2) was found to be thermally unstable. Furthermore, the yield in the first stage was low and variable (24 to 48%) impacting the overall cost and throughput. The AZ team developed a scalable, potential commercial route to AZD1981 by use of a Semmler-Wolff aromatisation of **2** to **3** to prepare the indole in 51% overall yield (Scheme 8.3).[11]

With the preferred route identified, process development moves on to delivery of a robust route of manufacture that provides consistent delivery of API of defined product quality (Control; next section) to ensure drug product efficacy and safety for the patient. In addition, development aims to minimise the overall impact (Environmental and Economic) through reagent and solvent selection, followed by optimisation to improve stage yields and the efficiency of materials used. Typically, organic solvents comprise the largest amount of material used in a manufacturing route[16] and the impact of their use can be minimised by careful process design to enable solvent

Table 8.2 Alignment of the SELECT criteria[13] with the 12 principles of green chemistry.[14]

Criterion	Description	12 Principles
Safety	Each stage should be safe to run on scale and the materials used should have minimal impact to human health.	3, 12
Environmental	The route should produce minimal waste and materials used should not be harmful to the environment.	1, 2, 5, 6, 8, 10
Legal	The route should be operable without infringement of intellectual property. Materials used should comply with existing regulations.	1, 3
Economic	The route should meet the development budget and ultimately the commercial cost of goods target.	2, 3, 4, 5, 6, 7, 8, 9, 11
Control	Control of operating parameters should be understood for each stage, leading ultimately to control of API quality.	11
Throughput	The availability and lead time of raw materials. The time taken to manufacture the API.	2, 5, 6, 7, 8, 9, 12

recovery where it is environmentally favourable to do so. Incineration, with energy recovery, of solvents with a low environmental impact of formation (*e.g.* diethyl ether, ethanol, hexane, heptane, methanol) has been shown to have a lower impact than distillation, whereas the converse holds for solvents with a higher environmental impact of formation (*e.g.* acetic acid, acetone, cyclohexane, ethyl acetate).[17] However, a recovered solvent needs to meet defined quality criteria if it is to be reused in API manufacture.[18] The route shown in Scheme 8.3 has a PMI of 117 $kg\,kg^{-1}$ API, which is comparable to the median PMI reported for an API in mid-phase development.[16] Consequently, it is reasonable to assume that continued process development would approximately halve the PMI at launch.[16]

If the route shown in Scheme 8.3 was filed as the registered synthesis, the stages shown would be run in accordance with current good manufacturing practice (cGMP) to ensure product quality.[19] The starting materials are referred to as the registered starting materials. In 2011, the European Directorate for the Quality of Medicines and Healthcare (EDQM) cited inappropriate proposal of registered starting materials as second in the top ten deficiencies in regulatory submissions. Their observation is that companies are proposing starting materials of significant complexity with only one or two registered stages to API.[20] EDQM does not regard this as

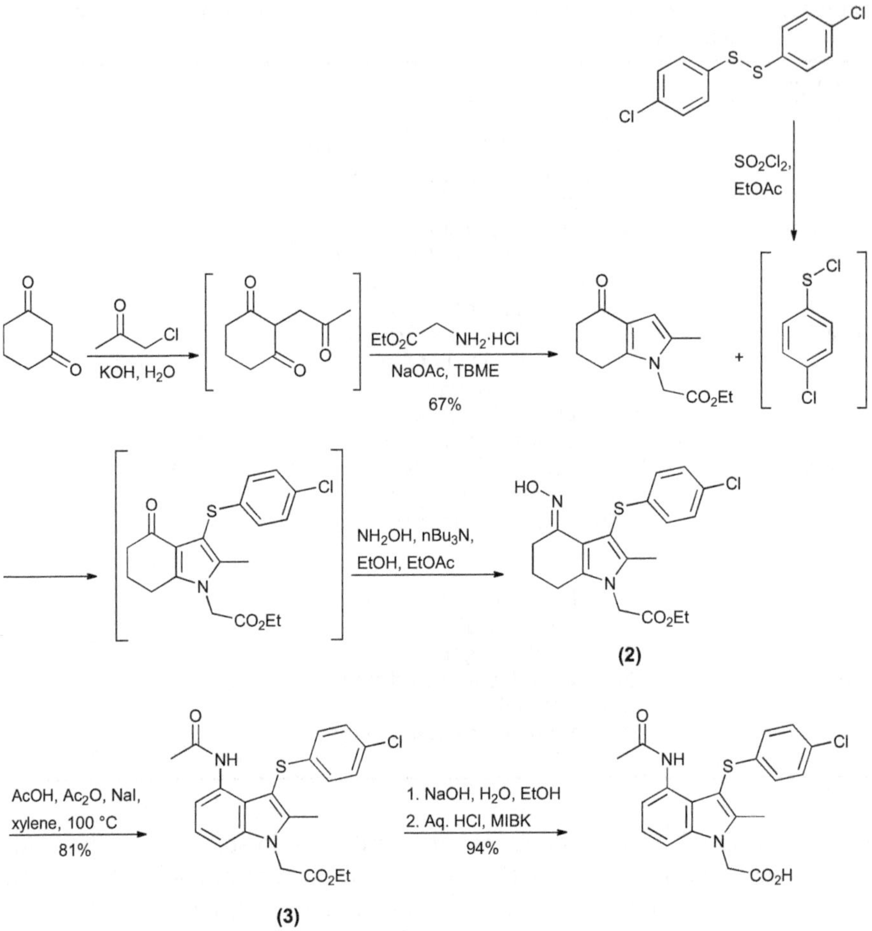

Scheme 8.3 A potential commercial route to AZD1981.[11]

providing sufficient information for them to assess the assurance of product quality.

The development of more efficient synthetic methodologies, which provide the ability to achieve specific bond formation in a single chemical transformation, replacing multiple stages of functional group manipulation (*e.g.* aromatic amination, C–H activation), has reduced the lifecycle impact of API manufacture. However, a consequence of this, and the need for sufficient stages to provide assurance of quality control, is that some registered starting materials are much closer to bulk commodity chemicals. These may be produced to a lower specification and require additional purification stages before use. Insufficient information to demonstrate understanding of proposed starting material specifications were the 4th and 5th deficiencies of regulatory submissions as cited by EDQM.[20] Principles for the selection of starting materials are outlined in ICH Q11.[21]

8.4 Quality Assurance

An API has to be consistently manufactured to high quality and the Chemistry Manufacturing and Controls (CMC) sections of a regulatory submission provide the summary for quality assurance of the drug product. This is achieved through a combination of product testing to meet minimum defined standards and process understanding, as discussed in Section 8.5.

8.4.1 Control of Process-related Impurities

The safety studies conducted before and during clinical trials enable dosing to patients over increasing periods of time. The quality of the API used in those initial studies defines the levels of the process-related impurities qualified by the safety assessment studies and helps define the specification of the API as it progresses through development.

Organic impurities $\geq$0.05% should be reported and those $\geq$0.10% (or 1 mg daily intake, whichever is lower) identified for a maximum daily dose $\leq$2 g (ICH Q3A).[21]

A significant activity during process development is developing an understanding of the origin, downstream purging (fate) and control of organic impurities. A change of route will need to produce API of comparable, or better, quality or contain new impurities below the ICH qualification threshold (0.15%). This may require additional purification stages, with concomitant waste generation. Alternatively, new impurities above the threshold will require qualification in an appropriate safety study. Typically the team will aim to use final route material in the long term safety and two year oncogenicity studies required to support a medicine for treatment of a chronic condition.

8.4.2 Residual Solvents

ICH Q3C(R5)[21] provides guidance on the permitted daily exposure (PDE) of common solvents, which translates to a residual solvent specification in the API, based upon the daily dose. Solvents are grouped into classes that broadly mirror the ranking of solvents in solvent selection guidance, as summarised in Table 8.3.[22]

A number of class 2 and 3 solvents (*e.g.* acetone, toluene, ethanol, isopropanol) may be contaminated with class 1 solvents (*e.g.* benzene), which will need to be controlled in the API; the absence of discussion for potential API contamination by a class 1 solvent was cited as the 6th common deficiency by EDQM.[20]

Careful solvent and reagent selection (*c.f.* Chapter 2: Tools for Facilitating More Sustainable Medicinal Chemistry) helps drive the use of lower toxicity solvents, which will require less stringent control. However, one challenge to the adoption of new, potentially greener, methodology is the absence of toxicology data for newer solvents that may have fewer environmental or

Table 8.3 ICH Q3C Classification of solvents.[21]

Class	Description	Example (concentration limit)	Solvent guide ranking
1	Solvents to be avoided (known or strongly suspected human carcinogens, known environmental hazards)	Benzene (2 ppm), carbon tetrachloride (4 ppm)	Not to be used
2	Solvents to be limited [non-genotoxic animal carcinogens, suspected of irreversible toxicities (*e.g.* teratogenicity) or suspected of significant but reversible toxicity]	Acetonitrile (410 ppm), dichloromethane (600 ppm), hexane (290 ppm), pyridine (200 ppm)	Red/amber
3	Solvents with low toxic potential to humans	Ethyl acetate, isopropanol[a]	Green

[a]Class 3 solvents are controlled to levels normally accepted in pharmaceuticals (typically not greater than 0.5%, 5000 ppm).

safety issues. In the absence of guidance, such a solvent has to be considered as a general impurity and the residual amount justified to the regulators. Generally, this will be to a tighter specification than a class 3 solvent, which can result in greater energy consumption through extended drying times, or in extreme cases displacement washing with a classified solvent. A PDE can only be calculated upon completion of safety testing, which will be conducted to enable registration of a chemical under the European REACH legislation if a solvent is to be commercialised.[23] However, the testing is extensive and can take a number of years to complete, giving a period of uncertainty. Furthermore, additional data can change the view of a solvent over time. For example, the recommended PDE for tetrahydrofuran was reduced from 121 mg day^{-1} to 7.2 mg day^{-1} (concentration limit 720 ppm), and moved from ICH Class 3 to Class 2 after *in vivo* carcinogenicity studies showed evidence of carcinogenic activity; THF did not give any positive findings during *in vitro* studies.[24]

8.4.3 Residual Metals

Control of residual metal content is outlined in ICH Q3D.[21] It is noteworthy that the metals currently most useful for chemical catalysis (Green Chemistry Principle 9), *e.g.* palladium and some platinum group metals (Rh, Ru, Ir), are subject to tight restriction because of known toxicity, or the assumption of near neighbour similarity in the absence of data. All four metals have an oral concentration limit of 10 ppm, reducing to 1 ppm *via* parenteral administration and 0.1 ppm *via* inhalation. Typically, delivery of API with low residual metal content can require additional processing and careful analysis to determine levels. Some of the more abundant metals suggested

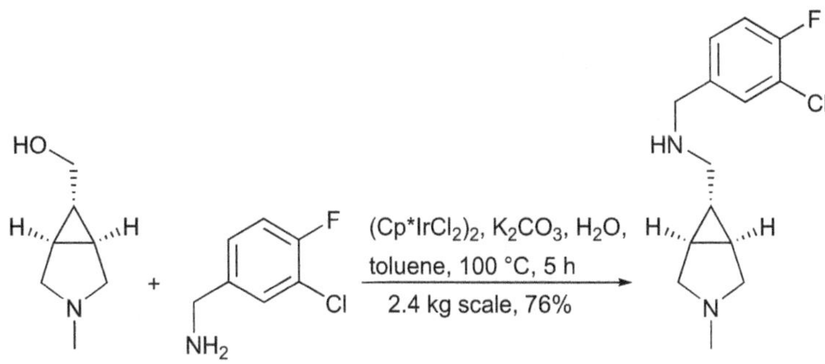

(4)

Scheme 8.4 Preparation of an intermediate to a GlyT1 inhibitor.[25]

as providing sustainable alternatives have higher oral concentration limits (*e.g.* copper; 300 ppm), whilst cobalt (5 ppm) and nickel (20 ppm) are subject to tight control because of known toxicity issues. The green chemistry aim to maximise the use and recovery of materials helps to deliver the required quality control. The development of reliable methods for the separation of homogeneous metals from process streams can be a challenge. For example, Berliner *et al.* screened 42 scavenging agents in their synthesis of an intermediate to a GlyT1 inhibitor (**4** in Scheme 8.4) to achieve acceptable iridium removal whilst maintaining a good yield.[25]

8.4.4 Transmissible Spongiform Encephalopathy

All materials used in the manufacture of drug products and product contact processing materials have to be assessed for the risk of contamination with transmissible spongiform encephalopathy (TSE).[26] Materials of the highest risk are those potentially derived from animal origin and include amino acids, some enzymes and charcoal supports for catalysts.

8.4.5 Genotoxic Impurities

The control of genotoxic impurities is an additional, specific aspect of quality control that has received considerable attention. Genotoxic impurities are defined as "DNA-reactive substances that have a potential for direct DNA damage".[27] The permitted level of residual genotoxic impurities has been the subject of some debate and proposals for "Assessment and Control of DNA Reactive (Mutagenic) Impurities in Pharmaceuticals to Limit Potential Carcinogenic Risk" are under advanced consideration through ICH M7.[5] Mutagenic materials give a positive Ames test. The permitted daily exposure of a mutagenic impurity is limited to 1.5 μg day^{-1} (the level associated with negligible risk to the patient being a theoretical excess cancer risk <1 in 100 000 over a lifetime of exposure, known as the threshold of toxicological concern). ICH M7 proposes to allow higher levels of exposure over a short duration. Consequently, the concentration in an API is dependent on the dose and duration of therapy. Demonstration of control requires specialised analytical techniques capable of detecting analytes at the low parts per million (ppm) level.

Application of green chemistry principles to synthetic route design should avoid the use of genotoxic materials and thus eliminate the need for testing. However, the current state of the synthetic art and, in particular, the high frequency of carbon–heteroatom bond formation in the synthesis of small molecule APIs means that the use of potentially genotoxic reagents can be quite common. In addition, combinations of some reagents and solvents can also lead to the formation of genotoxic impurities (*e.g.* sulfonic acids and alcoholic solvents). The combination of these factors means that the challenge of avoiding genotoxic materials is likely to be present for the foreseeable future. A number of publications have proposed a risk-based assessment of the likelihood of purging a genotoxic impurity to reduce the number analytes requiring detailed analysis.[28]

8.4.6 Control of Polymorphic Form and Stability

The clinical performance of a drug product can be highly dependent on the selected version (*e.g.* parent, salt, solvate or hydrate, co-crystal) and/ or its polymorphic form[29] and companies can have a team of specialist particle scientists and engineers working to ensure control of the final API forming stage. Analysis of the API's physical properties is undertaken to confirm polymorphic form, particle size, crystal habit and degree of solvation/hydration can form part of the API release testing regime for those attributes that are critical for performance. Bulk stability of the API is determined in the proposed packaging under defined combinations of temperature, humidity and light in order to understand their impact on drug substance quality and determine the retest period [ICH Q1A(R2)].[21]

8.5 Quality by Design

Historically, API quality assurance has been achieved through end-product testing to specification, as discussed in Section 8.4, and process development focussed on defining process parameter boundary conditions to ensure a successful outcome of the chemistry and the understanding of impurities. Current process development applies the concepts of risk-based Quality by Design, as summarised in ICH Q8 (R2): Pharmaceutical Development, Q9: Quality Risk Management, Q10: Pharmaceutical Quality System, Q11: Development and Manufacture of Drug Substances (Chemical Entities and Biotechnological/Biological Entities) and Q12: Technical and Regulatory Considerations for Pharmaceutical Product Lifecycle Management (proposed in 2014).[21]

Quality by Design encompasses a move away from quality control *via* end-product testing towards using a combination of risk management and scientific knowledge to develop an understanding of the linkage of process parameters to API and/or drug product critical quality attributes (CQAs, *e.g.* organic impurities, residual solvents). A combination of analytical testing, in-process analytics, and parameter control defines a control strategy for consistent delivery of APIs. A regulatory submission will typically include a design space developed through a series of studies exploring multi-variate parameters. The experimentation, even with the application of statistical methods, can be time consuming and will usually only be conducted once the synthetic route is defined.

An example of a quality by design approach is illustrated by the control strategy to limit the potential genotoxic aryl nitroso and hydroxylamine impurities (**5** and **6** in Scheme 8.5) arising from the reduction of an aromatic nitro-compound developed by Looker *et al.*[30]

The authors studied the potential process parameters that could have an impact upon completion of the reduction: catalyst loading; hydrogen pressure; reaction volume; reaction time; reaction temperature; distillation volume following catalyst removal; isolation temperature; number, volume and temperature of cake washes; and temperature and vacuum pressure of drying. The output of their development work enabled the definition of normal operating ranges within which the parameters could be controlled to enable successful completion of the reaction. The process knowledge was assessed by conducting a Failure Modes Effects Analysis (FMEA, see ICH Q9[21]), judging each parameter's impact on product quality or impact on another parameter. Low risk parameters were further assessed through range-finding studies. Three parameters were identified as high risk (catalyst loading, reaction temperature and time) and these were subjected to multi-variate analysis through statistical design of experiments.

The authors established a design space for palladium content (2.5–7.5 wt%) and temperature (15–35 °C) within which <0.03% of the by-products were formed after a minimum reaction time. They went on to examine the purge capability of the process to determine the levels of the impurities (0.34% nitroso and 0.83% hydroxylamine) that could be effectively purged during work-up to deliver impurity free product, and through spiking studies showed that 0.54% nitroso and 0.34% hydroxylamine in the intermediate were effectively purged through downstream chemistry to afford an API with neither impurity detected (<1 ppm).

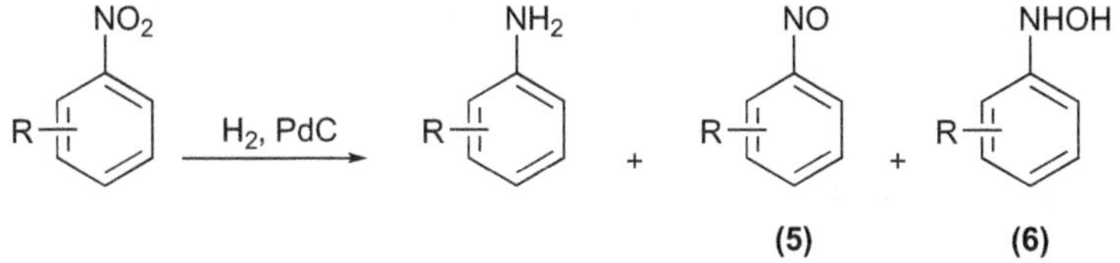

(5) (6)

Scheme 8.5 Reduction of an aromatic nitro compound.[30]

8.6 Compound Attrition

The decline in productivity of the pharmaceutical industry has been discussed at length in the literature.[6,31] For example, Scannell *et al.* reported that the number of drugs produced per $billion over time halved approximately every 9 years through to 2010, with the inflation adjusted R&D cost exceeding $1billion around the year 2000,[31a] although a report indicates that a corner may have been turned in 2013.[32]

The decline can potentially affect the environmental sustainability of APIs. For example, a trend towards an increasing number of molecules with greater lipophilic character[6,31c] generally results in APIs with poor aqueous solubility (Biopharmaceutical Classification II or IV),[33] which in turn can lead to poor oral bioavailability. A consequence of this is that additional processing may be required in the formulation in an attempt to improve absorption; *e.g.* physical size reduction through micronization or development of stabilised amorphous materials through spray drying. For example, if a poorly soluble drug with the median PMI for commercial products of 45 $kg\,kg^{-1}$ API[16] requires micronization with 90% yield, the PMI of the API in formulated product increases to 50 $kg\,kg^{-1}$, keeping in mind that is waste associated with the registered stages of the process and does not take into account the materials used in the manufacture of the starting materials.

High compound attrition, especially in the early phase of development, can drive companies to take a minimal development approach through clinical phases I and IIa, with delivery of API to support clinical development the only activity. If this approach is taken, the time available for the chemistry team to identify and implement their preferred manufacturing route is reduced; often this is driven by the need to manufacture API for long-term toxicology studies and phase III. One consequence can be progression of a sub-optimal route through to commercialisation resulting in a greater environmental impact.

There are approaches that can be taken to minimise the long term impact of this outcome through lifecycle management (ICH Q11 and Q12).[21] The development of second generation routes post launch is attractive in that the work is conducted on an asset that has been approved and so the effort is not wasted. However, it can take time to develop a route and considerable time to manage a change, in particular comparable drug performance has to be demonstrated and stability data has to be generated on API and drug product prepared by the new route. In addition, there will be a need to manage dual inventory during the changeover period. A publication by Dunn illustrates the improvements implementation of second generation routes can deliver and also discusses the time taken by various regulatory authorities to approve the change, which can be considerable.[34]

8.7 Conclusions

70 of the top 100 pharmaceutical products by sales in the USA in 2013 were small molecules or peptides and all contain at least one carbon–heteroatom

bond, 59 of the molecules contained at least one nitrogen atom, 42 were chiral and 41 contained at least one heterocyclic ring.[35] Furthermore, analysis of the chemical transformations used in API synthesis has shown little change in the last decade[36,37] so it is reasonable to assume that the current synthetic methodology challenges (carbon–heteroatom alkylation, acylation and carbon–carbon bond formation)[38] are likely to be the same in the near future. However, one might expect the shape, functionality and increased aqueous solubility of the reaction substrates to change in response to the challenges of reducing lipophilicity. These considerations emphasise the need to develop reliable, broadly applicable synthetic methodologies that will give discovery scientists confidence to use them by first intent.

Adoption of improved methodologies, and strategic application of technology, along with careful solvent selection and process design to ensure materials can be recycled, provides significant potential to reduce the overall impact of API manufacture in the years to come.

Acknowledgements

I wish to thank Helen Sneddon and Andrew Wells for their input and helpful discussions during the preparation of this chapter, which is prepared as part of CHEM21. CHEM21 is an Innovative Medicines Initiative Joint Undertaking under grant agreement no 115360, resources of which are composed of financial contribution from the European Union's Seventh Framework Programme (FP7/2007-2013) and EFPIA companies' in-kind contributions.[39]

References

1. S. Laufer, U. Holzgrabe and D. Steinhilber, *Angew. Chem. Int. Ed.*, 2013, **52**(15), 4072.
2. ICH Work Products, M4: The Common Technical Document, http://www.ich.org/products/ctd.html (accessed July 2015).
3. The International Conference on Harmonisation of Technical Requirements for Registration of Pharmaceuticals for Human Use (ICH), http://www.ich.org/ (accessed July 2015).
4. A. Giaquinto, L. Lindstrom, T. Tominaga, M. Ward, P. Doerr, L. Hunt and L. Rago, The Value and Benefits of ICH to Drug Regulatory Authorities — Advancing Harmonization for Better Health, http://www.ich.org/fileadmin/Public_Web_Site/News_room/C_Publications/ICH_20_anniversary_Value_Benefits_of_ICH_for_Regulators.pdf (accessed July 2015).
5. ICH Work Products, Multidisciplinary Guidelines, http://www.ich.org/products/guidelines/multidisciplinary/article/multidisciplinary-guidelines.html (accessed July 2015).
6. M. M. Hann and G. M. Keserü, *Nat. Rev. Drug Discovery*, 2012, **11**(5), 355.
7. R. Bonnert and R. Rasul, *International* WO 2004/106302, 2004.

8. T. Luker, R. Bonnert, S. Brough, A. R. Cook, M. R. Dickinson, I. Dougall, C. Logan, R. T. Mohammeda, S. Paine, H. J. Sanganee, C. Sargent, J. A. Schmidt, S. Teague and S. Thom, *Bioorg. Med. Chem. Lett.*, 2011, **21**(21), 6288.
9. C. Jimenez-Gonzalez, C. S. Ponder, Q. B. Broxterman and J. B. Manley, *Org. Process Res. Dev.*, 2011, **15**(4), 912.
10. C. J. Welch, D. W. Henderson, D. M. Tschaen and R. A. Miller, *Org. Process Res. Dev.*, 2009, **13**(3), 621.
11. M. Sulur, P. Sharma, R. Ramakrishnan, R. Naidu, E. Merifield, D. M. Gill, A. M. Clarke, C. Thomson, M. Butters, S. Bachu, C. H. Benison, N. Dokka, E. R. Fong, D. J. R. Hose, G. P. Howell, S. E. Mobberley, S. C. Morton, A. K. Mullen, J. Rapai and B. Tejas, *Org. Process Res. Dev.*, 2012, **16**(11), 1746.
12. R. Dach, J. J. Song, F. Roschangar, W. Samstag and C. H. Senanayake, *Org. Process Res. Dev.*, 2012, **16**(11), 1697, and reference 9 therein.
13. M. Butters, D. Catterick, A. Craig, A. Curzons, D. Dale, A. Gillmore, S. P. Green, I. Marzino, J.-P. Sherlock and W. White, *Chem. Rev.*, 2006, **106**(7), 3002.
14. P. T. Anastas and J. C. Warner, *Green Chemistry: Theory and Practice*, Oxford University Press Inc., New York, NY, 1998.
15. European Commission, Environment, The Industrial Emissions Directive, http://ec.europa.eu/environment/industry/stationary/ied/legislation.htm (accessed July 2015).
16. R. K. Henderson, J. Kindervater and J. B. Manley, Lessons learned through measuring green chemistry performance – The pharmaceutical experience (2007) http://www.acs.org/content/dam/acsorg/greenchemistry/industriainnovation/roundtable/lessons-learned-through-measuring-green-chemistry-performance-the-pharmaceutical-experience.pdf (accessed July 2015).
17. (a) C. Capello, U. Fischer and K. Hungerbühler, *Green Chem.*, 2007, **9**(9), 927; (b) A. Amelio, G. Genduso, S. Vreysen, P. Luis and B. Van der Bruggen, *Green Chem.*, 2014, **16**(6), 3045.
18. *WHO Drug Information*, 2011, **25**(3), 221.
19. U.S. Food and Drug Administration, Drug Applications and Current Good Manufacturing Practice (CGMP) Regulations, http://www.fda.gov/drugs/developmentapprovalprocess/manufacturing/ucm090016.htm (accessed July 2015) and ICH Q7A.[21]
20. European Directorate for the Quality of Medicines & Healthcare Top Ten Deficiencies New Applications for Certificates of Suitability (2011); http://www.edqm.eu/site/paphcep_12_15pdf-en-30964-2.html (accessed July 2015).
21. ICH Quality guidelines Q1 – Q12 can be found at: http://www.ich.org/products/guidelines/quality/article/quality-guidelines.html (accessed July 2015).
22. H. F. Sneddon, Tools for Facilitating More Sustainable Medicinal Chemistry, in *Green and Sustainable Medicinal Chemistry*, ed.

L. Summerton, H. F. Sneddon, L. C. Jones and J. H. Clark, Royal Society of Chemistry, Cambridge, ch. 2.

23. European Chemicals Agency REACH Regulations, http://echa.europa.eu/regulations/reach (accessed July 2015).
24. U.S. Food and Drug Administration Guidance (Drugs), PDE for tetrahydrofuran, http://www.fda.gov/downloads/Drugs/GuidanceComplianceRegulatoryInformation/Guidances/UCM216251.pdf (accessed July 2015).
25. M. A. Berliner, S. P. A. Dubant, T. Makowski, K. Ng, B. Sitter, C. Wager and Y. Zhang, *Org. Process Res. Dev.*, 2011, **15**(5), 1052, and supplementary information.
26. World Health Organization (WHO), Transmissible spongiform encephalopathies (TSE), http://www.who.int/bloodproducts/tse/en/ (accessed July 2015) and European Commission, Note for guidance on minimising the risk of transmitting animal spongiform encephalopathy agents via human and veterinary medicinal products (EMA/410/01 rev.3) (2011/C 73/01), http://www.ema.europa.eu/docs/en_GB/document_library/Scientific_guideline/2009/09/WC500003700.pdf (accessed July 2015).
27. EMEA, Committee for medicinal products for human use Guideline on the limits of genotoxic impurities: http://www.ema.europa.eu/docs/en_GB/document_library/Scientific_guideline/2009/09/WC500002903.pdf (accessed July 2015).
28. A. A. Teasdale, S. Fenner, A. Ray, A. Ford and A. Phillips, *Org. Process Res. Dev.*, 2010, **14**(4), 943; (b) A. Teasdale, D. Elder, S.-J. Chang, S. Wang, R. Thompson, N. Benz and I. H. Sanchez Flores, *Org. Process Res. Dev.*, 2013, **17**(2), 221; (c) D. Elder, G. Okafo and M. McGuire, *Org. Process Res. Dev.*, 2013, **17**(8), 1036; D. D. Elder, K. L. Facchine, J. N. Levy, R. Parsons, D. Ridge, L. Semo and A. Teasdale, *Org. Process Res. Dev.*, 2012, **16**(11), 1707.
29. (a) *Polymorphism in Pharmaceutical Solids*, ed. H. G. Brittain, Drugs and the Pharmaceutical Sciences, Marcel Dekker, Inc., New York, 1999, vol. 95; (b) *Pharmaceutical Salts and Co-crystals*, ed. J. Wouters and L. Quéré, RSC Drug Discovery Series No. 16, The Royal Society of Chemistry, Cambridge, 2012.
30. A. R. Looker, M. P. Ryan, J. Neubert-Langille and R. Naji, *Org. Process Res. Dev.*, 2010, **14**(4), 1032.
31. (a) J. W. Scannell, A. Blanckley, H. Boldon and B. Warrington, *Nat. Rev. Drug Discovery*, 2012, **11**(3), 191; (b) B. Munos, *Nat. Rev. Drug Discovery*, 2009, **8**(12), 959; (c) M. M. Hann, *MedChemComm,*, 2011, **2**(5), 349, and references cited therein.
32. U. Schulze, M. Baedeker, Y. T. Chen and D. Greber, *Nat. Rev. Drug Discovery*, 2014, **13**(5), 331.
33. The Biopharmaceutical Classification System: http://www.fda.gov/AboutFDA/CentersOffices/OfficeofMedicalProductsandTobacco/CDER/ucm128219.htm (accessed July 2015).

34. P. J. Dunn, *Green Chem.*, 2013, **15**(11), 3099.
35. (a) Njardarson research group Top 100 US prescription and brand name drugs, 2013 poster, http://cbc.arizona.edu/njardarson/group/top-pharmaceuticals-poster; (accessed July 2015) for the purposes of counting, combination therapies were counted as one and different formulations containing the same molecule were counted separately; (b) N. A. McGrath, B. Brichacek and J. T. Njardarson, *J. Chem. Ed.*, 2010, **87**(12), 1348.
36. J. S. Carey, D. Laffan, C. Thomson and M. T. Williams, *Org. Biomol. Chem.*, 2006, **4**(12), 2337.
37. J. Messinger, L. Otsomaa and S. Rasku, Medicinal chemistry – How "green" is our synthetic toolbox?, Chapter 9, Figure 9.5.
38. D. J. C. Constable, P. J. Dunn, J. D. Hayler, G. R. Humphrey, J. L. Leazer Jr., R. J. Linderman, K. Lorenz, J. Manley, B. A. Pearlman, A. Wells, A. Zaks and T. Y. Zhang, *Green Chem.*, 2007, **9**(5), 411.
39. The Innovative Medicines Initiative, www.imi.europa.eu. (accessed July 2015).

CHAPTER 9

Medicinal Chemistry: How "Green" is Our Synthetic Tool Box?

JOSEF MESSINGER,* LEENA OTSOMAA AND SIRPA RASKU

Orion Pharma, Medicinal Chemistry, Orionintie 1, Espoo, 02101 Finland
*Email: josef.messinger@orionpharma.com

9.1 Introduction

Medicinal chemistry is the art of designing and synthesizing compounds that interact with living organisms to obtain a certain response, leading to the cure of a deregulation in living organisms. To achieve this goal, medicinal chemists (sometimes also called pharmaceutical chemists) have different computational and synthetic tools available to support the different optimization cycles needed in drug discovery.[1,2]

The synthetic tools can be understood as 'all available chemical and biochemical transformations using classical batch or flow processes'. In this book chapter we will focus on chemical transformation.

The majority of drug discovery approaches in pharmaceutical companies still focus on target-based research activities in which a protein—believed to have a strong impact on disease—is modulated as required. Nevertheless, concepts like phenotypic approaches or polypharmacology are more frequently used nowadays owing to the understanding of disease and biological systems as naturally multi regulated systems. Phenotypic screening has been

RSC Green Chemistry No. 46
Green and Sustainable Medicinal Chemistry: Methods, Tools and Strategies for the 21st Century Pharmaceutical Industry
Edited by Louise Summerton, Helen F. Sneddon, Leonie C. Jones and James H. Clark

Published by the Royal Society of Chemistry, www.rsc.org

shown to be the most successful approach in discovery for first in class drugs during the years 1999–2008.[3]

9.2 From Hit to Candidate in Drug Discovery

In a classical setting, after a target is selected, the first round of research is dedicated to identifying hits. These hits will then be further explored in a "hit to lead" program and lead optimization, leading to a candidate for further development. During the chemical optimization, chemists have to combine many properties that are independent from each other into a single chemical structure. Properties are described by criteria that have to be fulfilled during the drug discovery process (Figure 9.1).

Each company has their own criteria set and decision points. Some generally accepted criteria can be defined as:

1. Hits are compounds that have confirmed activity on a target protein and are suitable for further optimization, also with regard to intellectual property (IP) rights.
2. Leads are compounds that have desired *in vitro* properties on ADME (absorption, distribution, metabolism, and excretion), are safe and patentable, and their activity is confirmed in first *in vivo* models.
3. Candidates are suitable for clinical development and are proven in disease-relevant *in vivo* models.

The terminology "pre-clinical development candidate" is also used regularly as the decision point after lead-optimization, before candidate selection, and often characterizes the necessity for regulatory toxicology studies, long lasting animal models and optimizing the synthetic route towards the first GMP (good manufacturing process) batch. During this

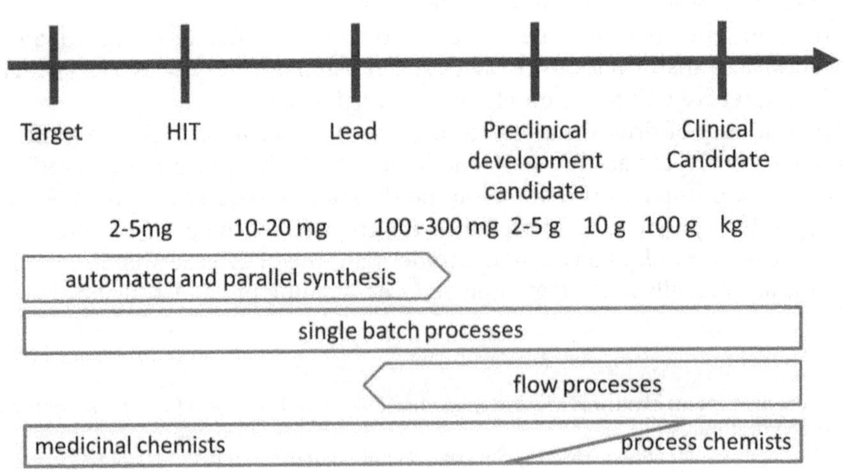

Figure 9.1 Scheme of a traditional research process in drug discovery.

phase, often the synthetic challenges are handed over from medicinal chemists to process chemists.

9.3 Multiparameter Space of Drug Discovery

Each of the discovery phases has different requirements, which vary a lot with regard to the amount of material needed, the purity of the compound, the physical form, *etc.* The most crucial point for designing the right properties into a structural motif with major input from medicinal chemistry is the "lead optimization" phase. In this phase, a highly flexible synthetic route is required with possibilities for incorporating bioactive groups to find the "optimal" compound in a multi parameter space defined by activity, selectivity, ADME, toxicology, patentability, drugability and further target-specific components (Figure 9.2). Thus, we will focus further on this phase to exemplify the synthetic tools and analyze them towards "greenness".

9.4 Lead Optimization Phase in Drug Discovery

As drug discovery is a highly competitive environment, "time" is an extremely important factor during the whole process. That is why all companies try to shorten their cycle time for optimization as much as possible in order to perform as many optimization cycles within a 1–2 year lead optimization phase. This is also the reason why, especially in the earlier phases of drug discovery, automated, combinatorial or parallel approaches play an important part (Figure 9.3).

Time plays an important role in the drug discovery process. Different synthetic strategies have to be used to obtain a diverse set of molecules. This is needed to explore the property space and to obtain appropriate answers (properties) during the optimisation cycle. This desire often leads to the concept of introducing important side chains and functionalities in the last

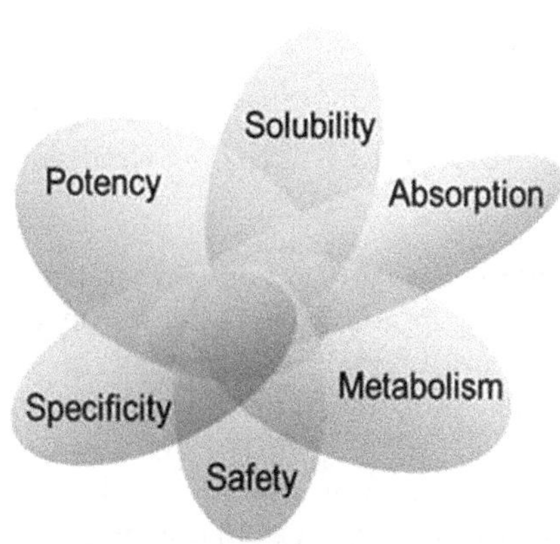

Figure 9.2 The optimization of active compounds towards drugs takes place in a multi parameter space.

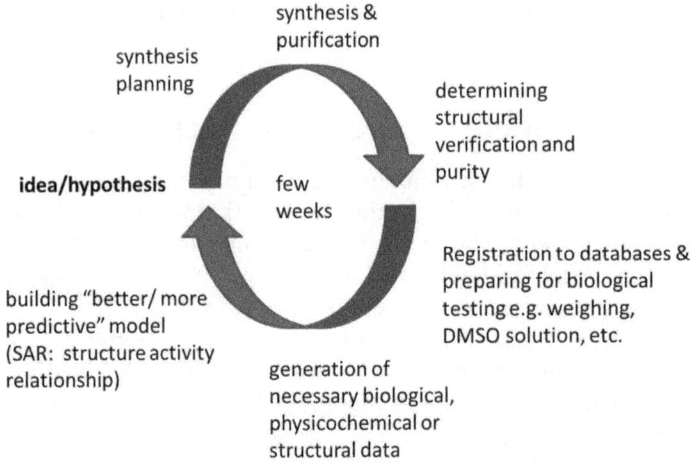

Figure 9.3 Important activities taking place during an optimization cycle.

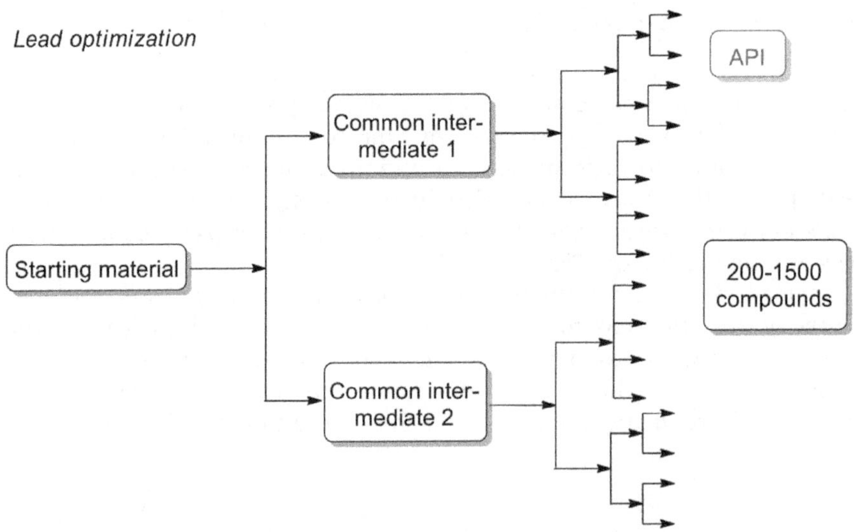

Figure 9.4 Depiction of strategy used during *e.g.* lead optimization (API = active pharmaceutical ingredient).

steps and working with key intermediates, common intermediates or building blocks as starting points, which as a consequence leads to less appropriate synthesis for individual molecules as the aim is not the optimal synthesis of a compound but is focused on speed to obtain the necessary set of compounds (Figure 9.4).

Often, during lead optimisation the required amount for performing *in vitro* testing is in the low mg region (<50 mg), which makes it necessary to perform resynthesis of a given compound on >100 mg scale. As also discussed in "Sustainable Practice in Medicinal Chemistry",[4] this should at least be considered for a first look at optimisation of the synthetic route—nevertheless, optimisation often only takes place if >1 g is needed for profiling of a selected compound.

9.5 Synthetic Tool Box and Reaction Analysis

The organic synthetic tool box is huge. In the 1990s, a chemist working in library synthesis might consider "March's Advanced Organic Chemistry" to represent the sum of what could be achieved. Now available in its 7th edition,[5] it covers some 1650 reactions/reaction types and 20 000 references. Despite these impressive numbers, it is only a small part of what is known compared to databases like REAXYS® or SciFinder®, which cover nearly 60 million reactions. It is not surprising that some believe that we have already explored all reactions[6] necessary to synthesize all possible bioactive compounds. However, many challenges and bottlenecks in chemistry exist when chemists in the pharmaceutical industry are striving to deliver their responsibilities in synthesizing and producing molecules.

The CHEM21 consortium[7]—which was set up to "make production of API's greener"—has compared the synthetic tools utilized by medicinal chemists and process chemists.

Interestingly, the analysis reflects what has been published in recent articles[8,9] showing that differences between discovery and development chemistry are small and focus on metal-catalyzed reactions. It is understandable that metal-based C–C and C–X formation reactions—like Suzuki, Buchwald, Hartwig, Stille, Sonogashira or metathesis reactions, to name just a few—are a more typical toolkit for medicinal chemists as they were mostly discovered and utilized during the last 20 years (Figure 9.5).

9.6 'Greenness' and Metrics

Before judging the 'greenness' of a synthetic tool, it is important to define and discuss some relevant issues.[4] There are three main areas to focus on (Figure 9.6):

- Safety and hazards of the reaction
- Reaction conditions (yield, workup, purification, solvent, and reaction temperature)
- Origin/availability of the reagents and waste (upstream and downstream considerations)

These considerations have been discussed in great detail and have been further elaborated and expanded with the development of the 12 principles of green chemistry.[10] Recently, Joseph P. Adams *et al.*[11] published a detailed analysis of reagents used in 15 transformations with regard to their safety and environmental impact. It is important to consider the selection of

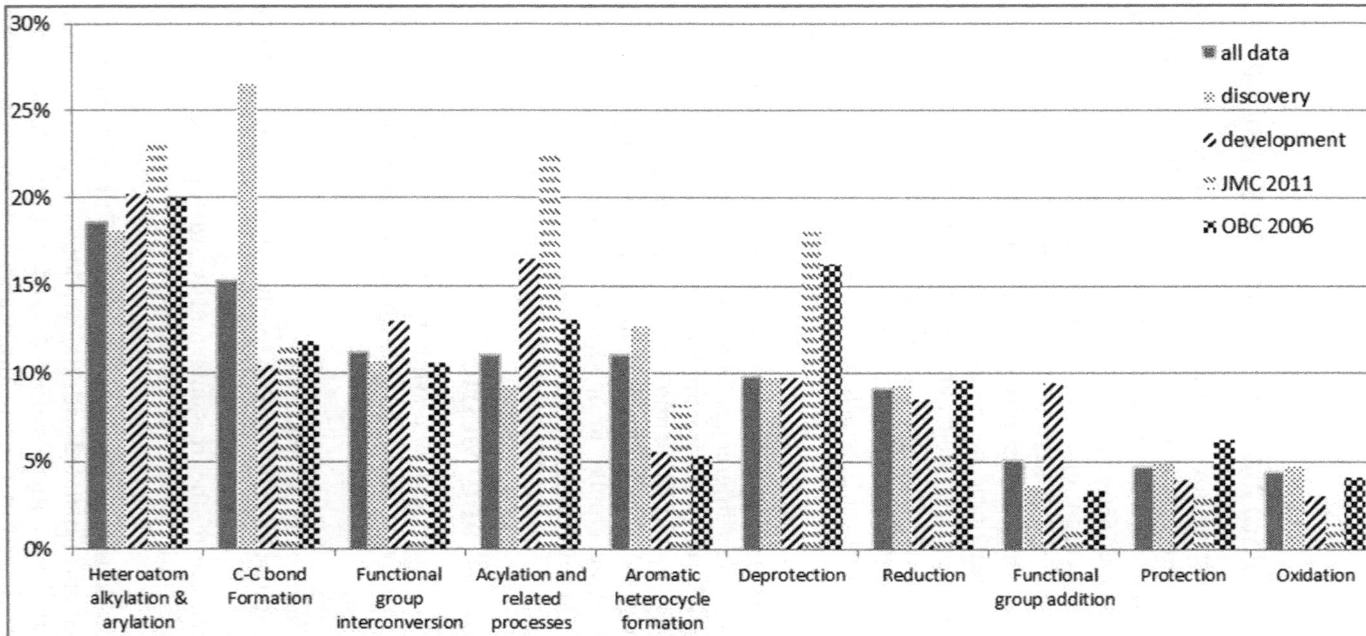

Figure 9.5 Graphic representation of different reaction types used during development and discovery phase within the IMI CHEM21 consortium compared to *The Journal of Medicinal Chemistry* and *Organic and Biomolecular Chemistry* articles.[8,9] Many thanks to all CHEM21 members who participated in collecting the data and to John Hayler and Andy Wells for compilation and analysis.

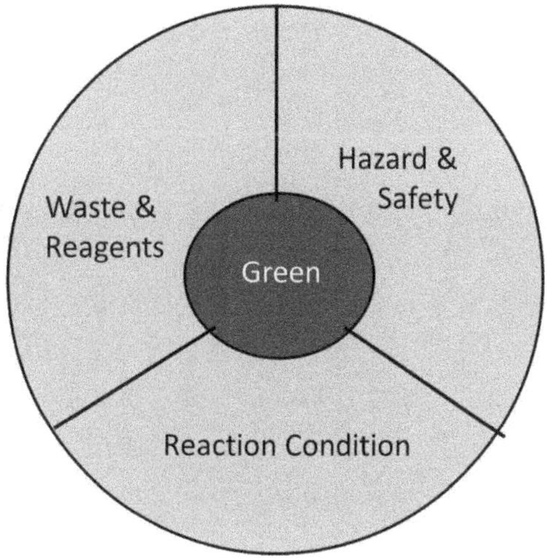

Figure 9.6 The three main areas for consideration when assessing the "greenness" of a reaction.

"greener" reagents; however, a holistic view on the overall reaction parameters is crucial when judging the overall environmental impact.

That is why within the CHEM21 consortium Prof. James Clark and his group (Rob McElroy, Louise Summerton, Andri Constandinou, University of York, York, UK) have generated a metrics 'tool kit' with complexity depending on the matureness of a synthetic sequence. In the case of small scale synthesis and early phases in discovery or development—which has been considered as phase 1 in the CHEM21 consortium—hard data like yield and mass intensity (MI)[12] as well as a flag assessment of solvents, work up *etc.* have been used (Figure 9.7).[13]

9.7 Magic Triangle: Choice of Synthetic Path

Medicinal chemists try to find a sustainable path for synthesis leading to a large variety of different compounds in the shortest possible timeframe, which can be visualized as a "magic triangle" (Figure 9.8). The diagram illustrates the constraints experienced by a medicinal chemist, where the desire to develop a 'green' low hazard route must be balanced against the need for rapid development/optimisation and a broadly applicable route allowing the preparation of a diverse range of analogues.

Issues related to the "magic triangle" may be exemplified by analyzing a recent *Journal of Medicinal Chemistry* paper by Laura Fumagalli *et al.* They describe in their paper the "Discovery of a potent and selective α1D-adrenoceptor antagonist - WB4101".[14] Their aim is to obtain more selective candidates towards adrenoceptor subtypes compared to a given starting point. With the synthesis of the dihydrobenzofuran fragment they have introduced a suitable intermediate to synthesise a variety of different

Parameter	Measures & Flags		
Yield	>89 %	70-89%	<70 %
MI/PMI (mass intensity/process mass intensity)			
Solvent	Use of recommended solvent	Use of problematic solvent	Use of hazardous or highly hazardous solvent
Catalyst/Enzyme	Catalyst/enzyme used or not needed	Stoichiometric quantities of catalyst needed	Excess of reagents used
	Catalyst is recovered	Catalyst is not recovered	
Critical Elements	All elements are widely available	Risk of depletion between 50-500 years	Risk of depletion between 5-50 years
Energy	Temperature range 0-70°C	Temperature range -20 - 140°C	Outside temperature range
	Reaction run 5°C or more below reflux		Reaction run at reflux
Reactor	Flow	Batch	
Work up	Quenching, filtration etc. (<140°C at atmospheric pressure)	Solvent exchange, quenching into aqueous solvent	Chromatography, high temperature distillation etc. (>140°C at atmospheric pressure), multiple crystallizations
Health & Safety	Highly explosive Explosive thermal runaway Toxic Long term toxic Environmental implications	H205, H220, H224 H241 H301, H311, H331 H341, H351, H361, H371, H373 H401, H412	H200, H201, H202, H203 H230,H240,H250 H300, H310, H330 H340, H350, H360, H370, H372 H400, H410, H411, H420
Chemicals of Concern			Subject to REACH authorization and restriction, on SINLIST
Availability	From >2 suppliers in >100 g quantities		
Applicability	Transformation is shown to work on wide range of reactants	Transformation is shown to work only on limited set of reactants	Transformation is shown to work only on selected reactants

Figure 9.7 Outline of metrics for phase 1 analysis of experimental data as agreed on by IMI CHEM21 consortium.[13] McElroy *et al. Green Chem.*, 2015, **17** (5), 3111–3121.

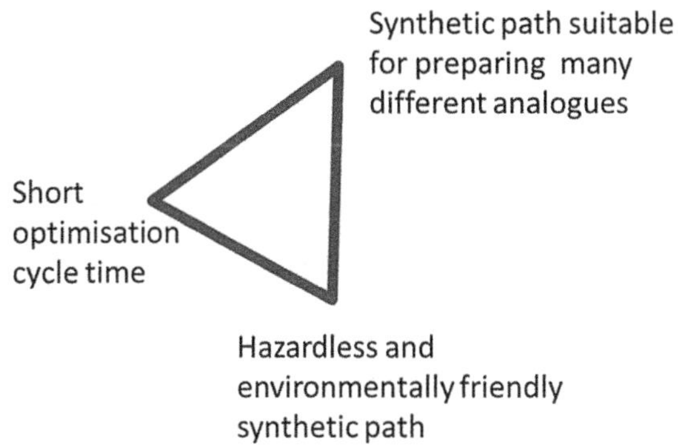

Figure 9.8 Magic triangle: choice of synthetic path depends on three main considerations.

compounds in a time-efficient way (Figure 9.9). The path towards dihydrobenzofuran fragment with 8 steps is not short, but suitable and time efficient for the purpose as in some cases no further purification was needed. However, the path had a significant drawback as step 6 had only 26% yield. Thus, the need for rapid method development and broad applicability needed to be weighed against the need for an environmentally friendly path.

9.8 Application of Green Chemistry Metrics

Overall, the scheme shown in Figure 9.9 is a 9 step synthesis including two crystallisations and 3 chromatographic purifications. The data provided has been used to calculate overall yield (6.3%) and mass intensity (MI = *ca.* 5000[15]). In total 11 different solvents were used. Interestingly, approximately 70% of the overall reaction mass can be attributed to workup and extraction procedures (and for ease of handling on a small scale). We can consider this to be a typical procedure for synthesizing derivatives. It also reflects the state of the art in a typical medicinal chemistry laboratory.

W. Cue Berkeley[16] used this and similar data for the calculation of MI (or PMI, process mass intensity) and waste produced during the different phases of drug discovery and came to the conclusion that overall early research (200–2000 t) and preclinical (150–1500 t) investigation produce more waste than in *e.g.* clinical phase 1 (40–400 t) development.

The example below highlights the question of whether parallel or microwave chemistry (MW) would have a significant effect on MI in the medicinal chemistry environment. As an example, comparison of a multimode MW and single mode MW have been used. The reaction studied was a Suzuki–Miyaura reaction with the aim to compare different catalysts and boron-sources like boronic acid, boronic acid ester and trifluoroborates (Figure 9.10).[17]

The focus of the reaction was high conversion and high recovery rate, which is why 2 equivalents of boron derivatives, approx. 10 mol% of Pd

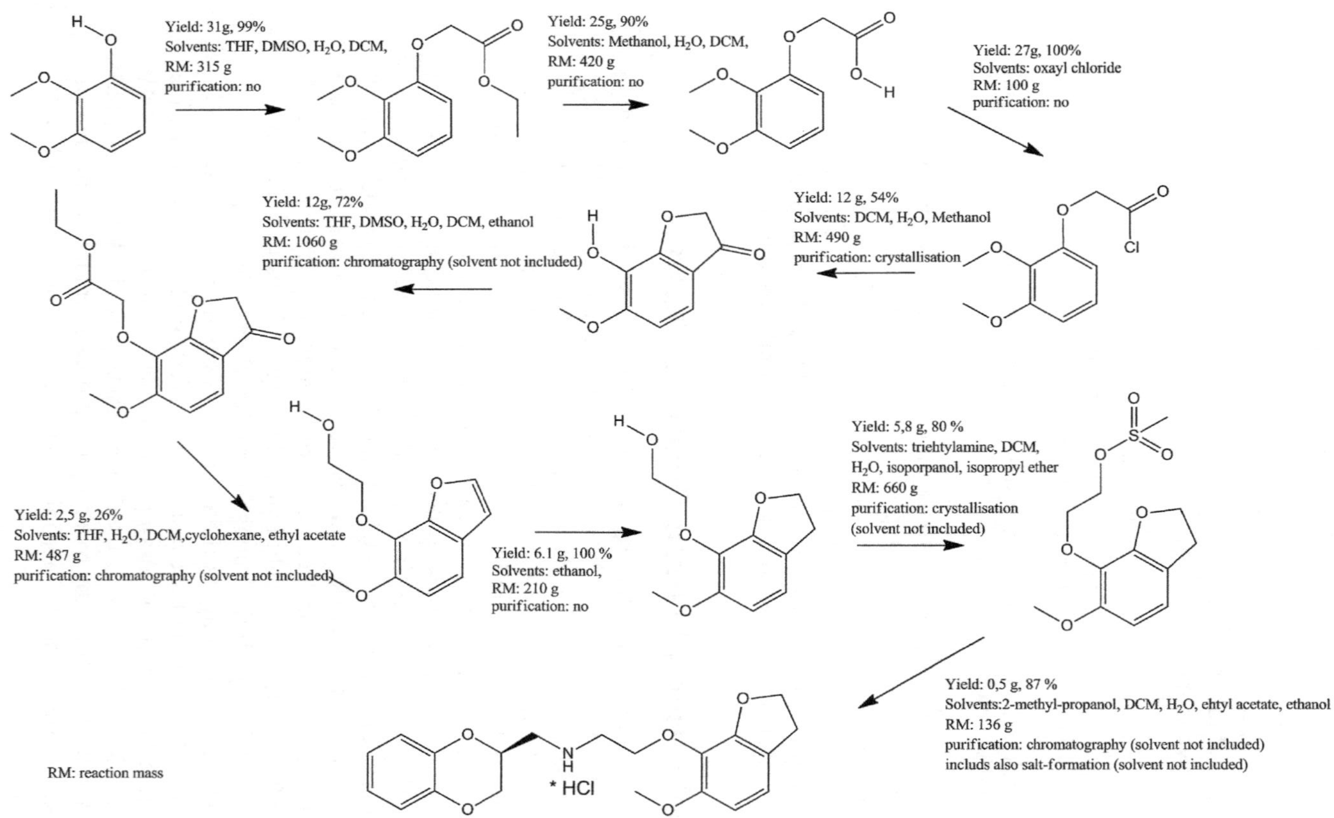

Figure 9.9 Synthesis of 2-[2-(2,6-dimethoxyphenoxy)ethyl]aminomethyl-1,4-benzodioxane including the information given in the paper to elucidate "greenness" of the synthetic path. THF: tetrahydrofuran; DMSO: dimethyl sulfoxide; DCM: dichloromethane. With permission from the ACS: L. Fumagalli, M. Pallavicini and R. Budriesi *et al.*, 6-Methoxy-7-benzofuranoxy and 6-methoxy-7-indolyloxy analogues of 2-[2-(2,6-Dimethoxyphenoxy)ethyl]aminomethyl-1,4-benzodioxane (WB4101): Discovery of a potent and selective α1D-adrenoceptor antagonist. *J. Med. Chem.*, 2013, **56** (16), 6402–6412.[14]

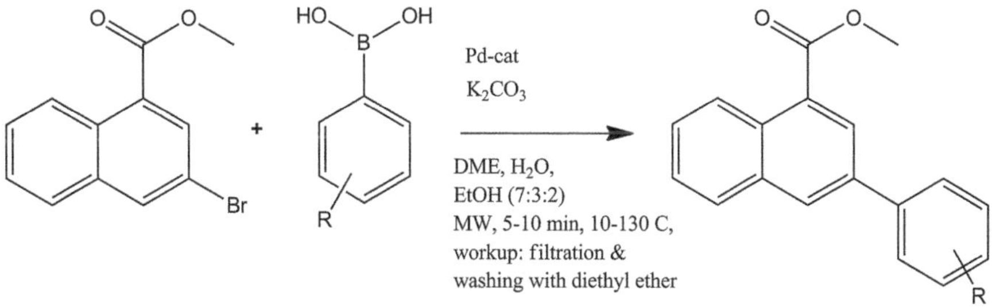

Figure 9.10 Suzuki–Miyaura reaction as performed by Uwe Schön *et al.* DME: dimethoxyethane, EtOH: ethanol.
With permission from Elsevier, U. Schön, J. Messinger, S. Eichner and A. Kirschning, Comparison of monomode and multimode microwave equipment in Suzuki–Miyaura reactions—en route to high throughput parallel synthesis under microwave conditions. *Tetrahedron Lett.*, 2008, **49**, 3204–3207.[17]

source and additional washing with diethyl ether after rinsing through Celite® has been used. This leads to an MI of approx. 200 (approx. 50 mg yield and total amount of reagents and solvents added to the reaction and during work up of close to 11 g). Compared to the reaction above with an average MI of 500 per step and MI of 72 in the last step, we see that the parallel approach is in a similar MI range. In total, 4 solvents have been used, including dimethoxyethane and diethyl ether. The isolated yield after purification *via* flash chromatography was >85% depending on the example.

In summary, automated methodologies may help to speed up the process but have more or less no effect on the other parameters of the "magic triangle".

9.9 How to Decrease MI and Amount of Waste?

The analysis of the two examples above raises further questions: "What can we do in research laboratories to decrease MI and the amount of waste? Do we have to use so many solvents, especially non-favourable solvents?"

The high mass intensity during research is generally a result of handling and purification procedures. Due to the fact that often reactions are performed on mg scale, higher dilution and more washing cycles are needed to ensure high recovery rates. Nevertheless, the use of *e.g.* micro reaction vessels (*e.g.* provided by Eppendorf) or microtiter plates, phase separation *via* centrifuges or solid-phase extraction for extractions or concentration of products (sometimes also referred to as catch and release techniques) is not yet broadly used to overcome at least partly the pitfalls of *e.g.* round bottom flasks or separation funnels. As handling has, so far, not been widely regarded as an issue, few research activities are focused on these technical aspects comparing different techniques with regard to "greener" handling.

A further issue is that, in general, synthetic routes in medicinal chemistry have not been optimized for a particular target molecule, which results in low yielding steps. *E.g.* doubling the yield in step 6 in case of the synthesis of dihydrobenzofuran in the example shown in Figure 9.9 would reduce the MI from 5000 to approximately 1500! People are still using reactions with mediocre yields despite all our experience, knowledge, and access to databases and reaction modelling tools that one has in hand nowadays. This indicates that we are still missing synthetic tools for certain reactions or even a better understanding of certain reaction types.

One reason may be that the implementation of new reaction technology is time consuming and not always a quick success story. In the article "Sustainable Practice in Medicinal Chemistry: Current State and Future Direction",[4] C–H activation was highlighted with a recent paper by Baran's group.[18] In Baran's group, they used Zn-reagents to introduce CF3-units to N-heteroaryl systems (Figure 9.11). A quick test-reaction in our hands led to approximately 7% yield, which is partly due to the fact that the final product is highly volatile but also to the fact that the best solvents and totally optimized reaction conditions focusing on green metrics are still missing and need to be optimized carefully with a larger set of substrates.[19]

A further way to reduce waste and lower MI during a reaction sequence of course is related to the number of steps. Having learned that isolation, purification and characterization of each intermediate are important, we rarely use telescoping or "one pot" reactions when synthesizing new compounds, which would also have a huge impact on the greenness of the reaction sequences. Another reason why people are hesitating in considering telescoping or "one pot" reactions is the choice of solvents. The main reason for this can be found in the tendency to just follow the literature method without reflecting further on why and which sort of solvents shall be used.[20] Furthermore, for the time being, chromatographic purification—especially

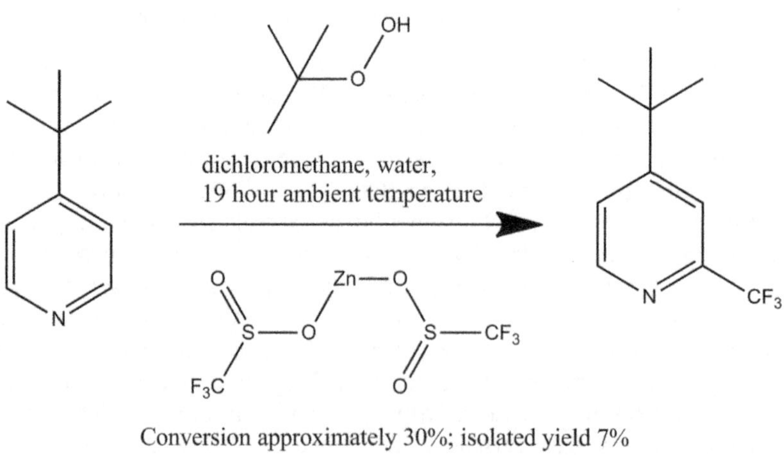

Figure 9.11 C–H activation using Zn-reagent analogues in Baran's reactions schema has been used in a test reaction for setting up "new chemistry" techniques.[19]

the widespread and growing use of preparative HPLC—cannot be avoided. Perhaps a reevaluation of the selectivity triangle for flash chromatography, taking into consideration the "greenness" of solvent, is necessary.[21]

Certainly, in the case of resynthesis of >1 g of a target molecule, it is necessary to consider the environmental impact of reactions and either use one's own experience gathered during process chemistry evaluation of projects or involve process chemists to give advice.

9.10 Energy Consumption

A final point to be discussed is energy consumption. In the case of the reaction itself, this is associated with reaction temperature, distillation procedure *etc.*, which are all important factors. However, the use of fume hoods and air conditioning in laboratories are the largest contributors to overall consumption of energy. That means that the proper handling of the ventilation systems, also during off times, has the greatest impact on the energy consumption in research labs.

9.11 How "Green" is Our Synthetic Tool Box?

Estimations as outlined above show that overall environmental impact of research laboratories is significant. The impact of an individual reaction might seem to be negligible but the sum of all our synthetic efforts cannot be ignored. The most significant impact is the choice and amount of solvents used—here we can learn a lot from our colleagues in process chemistry. On the other hand, we need to explore reaction, work up and purification techniques further and compare the new technologies based on an overall environmental footprint. As cycle time and diversity of reactions play a crucial role in early drug discovery, synthetic tools need to be available that have been explored for a broad set of reagents and optimized with regard to solvents considered to be 'non-problematic'.

9.12 Summary

1. Medicinal chemistry is the art of designing and synthesizing compounds that interact with a living organism to obtain a certain response leading to curing a deregulation in a living organism.
2. The phases of drug discovery are target finding, hit finding, hit to lead program or lead finding, lead optimization (sometimes pre-clinical candidate selection), and finally clinical candidate selection.
3. Important activities during *e.g.* the lead optimization cycle are starting hypothesis and structural design, synthesis planning, synthesis and purification, structural verification and purity check, registration to database preparing for biological testing, generation of necessary biological and physicochemical data, verifying hypothesis and building improved structure–activity model.

4. Medicinal chemists try to find a sustainable path for synthesis leading to a large variety of different compounds in a shortest possible timeframe.
5. The amount of waste produced in research laboratories around the world is significant and comparable to the amount of waste produced during the production of the first clinical trials worldwide.
6. The choice and amount of solvents used has the most significant impact on the greenness of a reaction, as well as the yield.

References

1. For further reading on Medicinal Chemistry please look to the large number of available text books; *e.g.* L. P. Graham, *An Introduction to Medicinal Chemistry*, Oxford University Press, 2009.
2. For further reading on Medicinal Chemistry please look at the large number of available text books, *e.g.* D. Steinhilber, M. Schubert-Zsilavecz and H. J. Roth, *Medizinische Chemie*, Deutscher Apotheker Verlag, Stuttgart, 2010.
3. D. C. Swinney and J. Anthony, *Nat. Rev. Drug Discovery*, 2011, **10**, 507.
4. M. C. Bryan, B. Dillon, L. G. Hamann, G. J. Hughes, M. E. Kopach, E. A. Peterson, M. Pourashraf, I. Raheem, P. Richardson and D. Richter, *J. Med. Chem.*, 2013, **56**(15), 6007.
5. M. B. Smith and J. March, *March's Advanced Organic Chemistry: Reactions, Mechanisms, and Structure*, John Wiley & Sons, 7th edn, 2013.
6. M. Hartenfeller, M. Eberle, P. Meier, C. Nieto-Oberhuber, K.-H. Altmann, G. Scheiner, E. Jacoby and S. Renner, *J. Chem. Inf. Model.*, 2012, **52**(5), 1167.
7. IMI consortium: "CHEM21 – Chemical Manufacturing Methods for the 21st Century Pharmaceutical Industries" has received funding from the Innovative Medicines Initiative Joint Undertaking under grant agreement n°115360, resources of which are composed of financial contribution from the European Union's Seventh Framework Programme (FP7/2007-2013) and EFPIA companies' in kind contribution. For further details see: http://www.chem21.eu/ (accessed November 14th, 2013).
8. S. D. Roughley and A. M. Jordan, *J. Med. Chem.*, 2011, **54**, 3451.
9. J. S. Carey, D. Laffan, C. Thomson and M. T. Williams, *Org. Biomol. Chem.*, 2006, **4**, 2337.
10. Green Chemistry Pocket Guides; http://bit.ly/ACSGCIPocketGuides (accessed November 14th, 2013).
11. J. P. Adams, C. M. Alder, I. Adams, P. Joseph, C. M. Alder, I. Andrews, A. M. Bullion, M. Campbell-Crawford, M. G. Darcy, J. D. Hayler, R. K. Henderson, C. A. Oare, I. Pendrak, A. M. Redman, L. E. Shuster, H. F. Sneddon and M. D. Walker, *Green Chem.*, 2013, **15**(6), 1542.
12. See Chapter 4 in this book by L. Summerton, A. Constandinou: Beyond Mass-based Metrics: Evaluating the Greenness of Your Reaction.
13. C. R. McElroy, A. Constantinou, L. C. Jones, L. Summerton and J. H. Clark, *Green Chem.*, 2015, **17**(5), 3111, The original table can be found on the CHEM21 web page: http://www.chem21.eu/project/metrics-toolkit/.

14. L. Fumagalli, M. Pallavicini, R. Budriesi, C. Bolchi, M. Canovi, A. Chiarini, G. Chiodini, M. Gobbi, P. Laurino, M. Micucci, V. Straniero and E. Valoti, *J. Med. Chem.*, 2013, **56**(16), 6402.
15. Comments on the calculation of MI: Solvents used for chromatographic purifications or crystallisations have been neglected in the calculation as they have not been mentioned in the experimental description. As from step to step not all material has been used or a reaction has been performed several times an appropriate, scaling has been done to calculate MI for the reaction sequence.
16. W. C. Berkeley, *Green Techniques for Organic Synthesis and Medicinal Chemistry*, ed. W. Zhang and W. C. Berkley Jr., John Wiley & Sons, 1st edn, 2012, vol. 20, p. 553. DOI: 10.1002/9780470711828.ch20.
17. U. Schön, J. Messinger, S. Eichner and A. Kirschning, *Tetrahedron Lett.*, 2008, **49**, 3204.
18. Y. Fujiwara, J. A. Dixon, F. O'Hara, E. D. Funder, D. D. Dixon, R. A. Rodriguez, R. D. Baxter, B. Herle, N. Sach, M. R. Collins, Y. Ishihara and P. S. Baran, *Nature*, 2012, **492**, 95.
19. The investigations were carried out as a student project at Orion Pharma as part of work for CHEM21-IMI project in WP2. Reaction details: A solution of 4-tert-butyl pyridine (1.5 mmol) and bis(trifluoromethylsulfinyloxy)zinc (4.50 mmol) in dichloromethane (6 mL) and water (2.4 mL) was cooled to 0 °C, followed by slow addition of tert-butyl hydroperoxide (7.50 mmol) under vigorous stirring. The reaction mixture was stirred at ambient temperature for 5 hours. Then an additional portion of reagents bis(trifluoromethylsulfinyloxy)zinc (4.50 mmol) and tert-butyl hydroperoxide (7.50 mmol) were added to the reaction mixture and the mixture was stirred for an additional 14 hours at ambient temperature. LC-MS analysis showed approximately 30% conversion. The reaction was worked up by adding dichloromethane (15 mL), saturated $NaHCO_3$ solution (15 mL) and 2–3 mL of water (to make phase separation easier). Layers were separated and the aqueous phase was extracted with dichloromethane (3×15 mL). The combined organic layers were washed once with saturated $NaHCO_3$ solution and concentrated carefully under reduced pressure. The crude product (730 mg, including some remaining dichloromethane) was purified using flash chromatography (diethyl ether/pentane gradient from 5 to 10% diethyl ether) yielding 21 mg of 4-tert-butyl-2-(trifluoromethyl)pyridine (0.103 mmol, 7% yield).
20. For further reading please refer to Chapter 2 in this book by H. Sneddon, Tools for Facilitating More Sustainable Medicinal Chemistry.
21. M. F. Vitha and A. R. Johnson, *J. Chromatogr.*, 2011, **1218**, 556.

CHAPTER 10

Design of Experiments (DoE) for Greener Medicinal Chemistry

JONATHAN D. MOSELEY

CatSci Ltd, CBTC2, Capital Business Park, Wentloog,
Cardiff CF3 2PX, UK
Email: jonathan.moseley@catsci.com

10.1 Introduction

Experimental design, or Design of Experiments (DoE) as it is more generally known, is a structured and efficient approach to experimentation that employs statistical techniques to investigate potentially significant factors and determine their cause-and-effect relationship on the outcome of an experiment.[1–3] Where there is a relationship between the experimental parameters (factors) and the results, the correlation can be discerned and quantified. Therefore, DoE can be usefully employed on any process for which the inputs (factors or parameters) can be controlled, and the outputs (results) can be measured in a statistically valid manner.

Consequently, DoE has been widely employed for many years and across many industries, sectors and disciplines, including more latterly the pharmaceutical industry. In this last case, it is most often employed in the development and scale-up of new processes for the manufacture of intermediates for potential new drugs (once the synthetic route is decided).[4,5] It has also often been used in the formulation processes required for the final

RSC Green Chemistry No. 46
Green and Sustainable Medicinal Chemistry: Methods, Tools and Strategies for the 21st Century Pharmaceutical Industry
Edited by Louise Summerton, Helen F. Sneddon, Leonie C. Jones and James H. Clark

Published by the Royal Society of Chemistry, www.rsc.org

active pharmaceutical ingredient (API),[6] and once the new drug is successfully launched, it may be used in the initial manufacturing processes to optimise the yield and quality further, and to reduce the manufacturing costs (although these studies are rarely published).

None of these activities hold much relevance to medicinal chemists, as they occur after their role is largely completed. It can be argued that DoE techniques are not relevant to early "discovery phase" activities, but are better suited to the optimisation of developing technologies and established industrial processes. Whilst it is true that DoE is not as widely used in academic departments, there are numerous examples of successful applications reported from disparate research laboratories, as this eclectic sample list of reviews exemplifies.[7–11] Although this topic has been somewhat treated before with drug discovery in mind,[12] the aim of this chapter is to convince the medicinal chemist of the value of DoE, *applied at the appropriate time*; explain briefly the DoE technique; and to provide guidance on practical considerations. A brief case study will conclude the chapter by way of illustration.

10.2 Why use DoE in Medicinal Chemistry?

The benefits of DoE have been frequently stated,[1–4,12] although even development chemists have needed persuading.[13] But many of these benefits appear of little relevance to the medicinal chemist who does not know if they are going to repeat the synthesis. So in what ways could DoE benefit medicinal chemistry, and thereby make it inherently "greener"? There are times when it may be appropriate for a medicinal chemistry project to seek an improved route, for example:

- Where a common core is decided upon early in the project, for which optimisation is necessary or would facilitate obtaining a viable supply of this core structure
- On a particularly difficult step where low yield or high volumes create a throughput bottleneck in the synthetic route
- For repeat supplies of a compound of interest in the early testing phase

For the medicinal chemist, all three reasons essentially mean making more material (intermediate or API) in a shorter time, with the emphasis on time. The medicinal chemist is not primarily interested in raw material cost, efficiency or optimisation, as they simply want to get new molecules tested for biological activity as quickly as possible. Of course, any improvements and efficiency gains in synthesis that the medicinal chemist develops during this period will be invaluable in both their own scale-up laboratories and to their process development colleagues who take over the eventual scale-up of promising candidates. This will be particularly useful to support early toxicology testing, which usually happens within the medicinal chemist's remit in a kilo-scale lab. It is often at this stage in route development, where processes are un-optimised, yields are low, and work-ups problematic, that material supply becomes an issue. Early application of DoE on gram scale syntheses will benefit itself, as well as at the 100–1000 g scale, where material supply often becomes critical against tight project deadlines.

So although the use of DoE on a medicinal chemistry project is likely to arise to overcome problems of material supply and to save time, it is the more efficient and optimised synthesis, and therefore the greener one, that will deliver this. Benefits are invariably cumulative because:

- Generally, optimised reactions give higher yields with fewer impurities, resulting in simpler and easier work-up processes
- Optimised reagent stoichiometries mean less in, hence less out, which also results in simpler and easier work-up processes
- Choice of more benign solvents may also result in easier work-ups, and will certainly be greener (see also Chapters 2 'Tools for Facilitating More Sustainable Medicinal Chemistry' and 3 'Renewable Solvent Selection in Medicinal Chemistry')

Note that usually it is the work-up and purification processes that generate poor metrics; high conversions to the desired product(s) will ease the purification process if troublesome impurities are not present or are much reduced. Although the investment of effort in optimisation when the demand for material supply is all-consuming appears to be the wrong priority, the more efficient, ultimately quicker (and greener) synthesis relies on developing reaction understanding as early as possible. DoE can be a great aid to deliver this. Of course, it remains a matter of judgement on which projects to invest effort, and at what stage of their development, but this is an eternal project management quandary.

10.3 Design of Experiments Explained

As already stated, DoE is a structured and efficient technique to uncover and quantify potentially significant factors and their cause-and-effect relationships on the outcome of a process (a reaction in this case). This is not the place to go into a detailed explanation of how DoE functions, as this has been well covered elsewhere.[1–3,13] However, a key advantage of its structured approach is in comparison to the chemist's traditional "one-variable-at-a-time" (OVAT) approach. DoE varies several factors at once, and can extract more data from a smaller set of experiments. Furthermore, OVAT is not only less efficient, but may fail to identify the best conditions possible in a given case (Figure 10.1).

The responses can be quantified and used to extrapolate to better outcomes (yield, selectivity, purity). Because multiple factors are varied in a rational manner, the interactions between them can also be identified and quantified, which again is not possible with OVAT. It is also often overlooked that because DoE uses rigorous statistical techniques, it can be used to

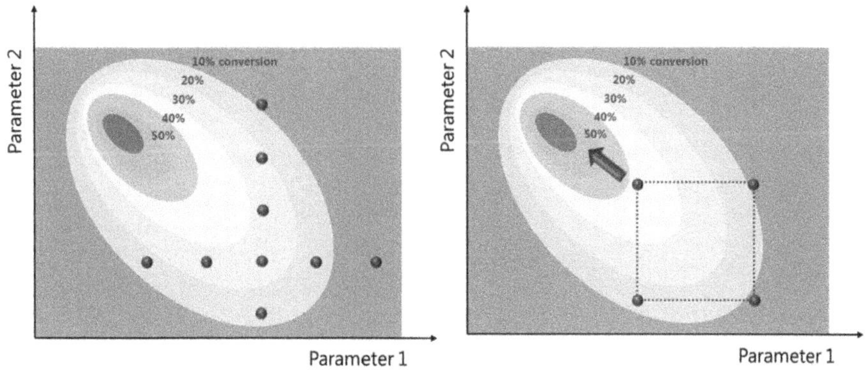

Figure 10.1 Comparison of OVAT (left) and DoE (right) techniques. Reproduced with permission from Figure 1 in ref. 14: J. D. Moseley and P. M. Murray, Ligand and solvent selection in challenging catalytic reactions, *J. Chem. Technol. Biotechnol.*, 2014, **89**, 623, Wiley. © 2014 Society of Chemical Industry.

determine the experimental error in the process—indeed, it is essential to determine the reproducibility to qualify the results of a DoE study. Once again, OVAT can offer no measure of this, and this may disguise the adverse influence of an uncontrolled factor until more rigorous experiments are conducted, which DoE would have uncovered much sooner.

For example, taking three factors and exploring their values at two levels, high and low, gives a standard cubic design (Figure 10.2). This requires 2^3 (*i.e.* 8) experiments for a full design. To determine experimental error and reproducibility, three centre points should also be performed. This is shown graphically in Figure 10.2, and in tabular form in Table 10.1.

As more factors are introduced, the number of experiments grows exponentially. Graphical representation beyond three factors in a cube is however difficult to represent (levels are less difficult). Fortunately, the software programmes available[15] can cope with what the human eye cannot visualise. Furthermore, experienced practitioners of DoE often run fractionated experimental designs when much larger experimental sets are required. In such cases, only a proportion of the planned experiments is performed in the first cycle, but the choices are made in a rational and statistically valid manner. It is in running fractionated designs that DoE *appears* to give more information than is possible, but this is not in fact the case. Indeed, some information on the more complex interactions may be lost due to confounding, which occurs when the main effects cannot be separated from the interaction effects due to too much fractionation. But it is rare for higher order interactions to be significant, so as long as the design is not too heavily fractionated, the main effects and the two factor interactions should be revealed and quantified.

This allows many more factors to be adequately explored, without having to perform a huge number of experiments.[16] However, the advent of

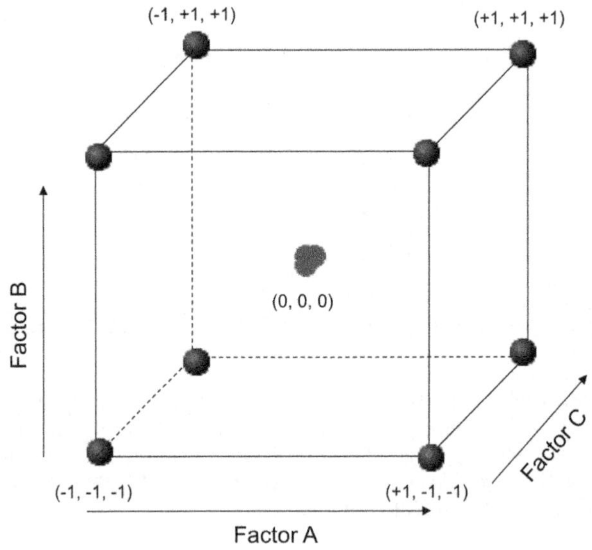

Figure 10.2 Graphical representation of a three factor, two level design. Reproduced with permission from Figure 4 in ref. 14: J. D. Moseley and P. M. Murray, Ligand and solvent selection in challenging catalytic reactions, *J. Chem. Technol. Biotechnol.*, 2014, **89**, 623, Wiley. © 2014 Society of Chemical Industry.

Table 10.1 Coding for a three factor, two level design with three centre points. Coding for the effect of interactions is also shown.

Run	Settings			Main Factor Coding			Interaction Factor Coding			
#	Factor A	Factor B	Factor C	A	B	C	AB	AC	BC	ABC
1	Low	Low	Low	−1	−1	−1	+1	+1	+1	−1
2	Low	Low	High	−1	−1	+1	+1	−1	−1	+1
3	Low	High	Low	−1	+1	−1	−1	+1	−1	+1
4	Low	High	High	−1	+1	+1	−1	−1	+1	−1
5	High	Low	Low	+1	−1	−1	−1	−1	+1	+1
6	High	Low	High	+1	−1	+1	−1	+1	−1	−1
7	High	High	Low	+1	+1	−1	+1	−1	−1	−1
8	High	High	High	+1	+1	+1	+1	+1	+1	+1
9–11	Mid	Mid	Mid	0	0	0	0	0	0	0

powerful automation for chemical synthesis and analysis does now make large experimental programmes possible in short times. A more limiting factor is likely to be starting material availability, especially if it needs to be synthesised. But if reactions can be easily performed in dilute solutions in small tubes or vials, or on small scale, then many more can be run. The key is to separate the reaction from the work-up (see below); if hyphenated analytical techniques can be applied to the reaction (*i.e.* those techniques combining chromatography for separation, with spectroscopy for identification/quantification, often LC-MS and GC-MS), then precious material can generate much more data. All the reactions can then be worked up together as a single batch, which is much more efficient, and knowledge from the DoE can then be applied to the next cycle.

Note that because DoE is planned, and an iterative process, further cycles build on the initial results—nothing is redundant. This could be by filling in the other experiments not performed in the first round of a fractional DoE; or by further investigating the design space in a particular region around promising results, again in a systematic manner. Right from the outset, DoE generates reaction understanding—it doesn't start from scratch all over again, as is common with OVAT. In the worst case, the chemist will be no worse off (because the material still has to be synthesised), and will know that the reaction cannot be improved as hoped.[17] In such circumstances, they either have to accept it, or find another route, but that is now a known limitation. If a significant factor has not been tested, this will be revealed by the statistics from the apparent lack of relationships between the factors and results, so further investigations are justified. More typically, the first round of a DoE will yield worthwhile immediate improvements in the reaction and indicate in what direction to adjust the factors for additional benefits. Even apparently "impossible" predictions can be realised with a little lateral thinking. One experimental design predicted that improved results could be achieved *above* the boiling point of the preferred solvent, so the researchers cleverly switched to a sealed tube microwave reactor,[18] technology with which most medicinal chemists will be familiar.

10.4 Practical Considerations

For those who have not used DoE before, knowing where to start is a common concern. But as when tackling any unfamiliar or new reaction, the medicinal chemist will be able to find a standard method in the literature, if they do not have one already. Typically, academic procedures will have been optimised by an OVAT approach, and may have been developed for other criteria. For example, many academic and medicinal chemistry reactions can be run at much higher concentrations than normally attempted. Therefore, even a cursory DoE is likely to yield quick improvements in the process.

One important point needs to be noted at the outset, however. The presentation of results in many academic publications often overlooks the difference between isolated yield and reaction conversion. The isolated yield (of the desired product(s)) is often taken as the direct outcome of the reaction parameters, when it is in fact a combination of both the reaction parameters and the work-up conditions. For the medicinal chemist, there may be some grounds for assuming this is the case; if a standard aqueous work-up and extraction leading into purification by chromatography *which isolates and quantifies all the products of the reaction is achieved*, then this will be the case. But this is an assumption, and a potentially misleading one at that. Accurate quantification by isolation of minor products is tedious at best and difficult otherwise; important minor stereoisomers may be lost or diminished in the work-up process on small scale.

Work-ups inevitably perturb the apparent reaction outcome to some extent, which may misrepresent or mask the important factors in a reaction, as some researchers have found initially.[19] It is far better if the conversion to all products in solution, including the ratio of any stereoisomers before isolation, can be quantified before work-up by a technique such as LC or GC-MS. Representative sampling is of course required, and slurries can be particularly problematic. Ideally, solution sampling will allow results to be related back to the reaction conditions; the work-up should be treated as a completely separate (*i.e.* derivative) process. But as noted, this does have advantages if multiple reactions can be worked up together to save time.

For the medicinal chemist at the start of a project which may not continue beyond 10 g, let alone 100 g, a DoE of 8–16 experiments (plus three centre-point replicates) is likely to identify the main factors and provide a significant improvement in yield, selectivity or quality. Probably only a few factors are likely to be important for a coarse reaction optimisation. A list of potential factors is shown in Table 10.2, but in practice it is likely to be limited to reagent stoichiometries and addition rates, temperature, time, concentration, and pH if relevant. Other factors only tend to become significant on scale-up and in more concentrated solutions, especially slurries, where factors such as agitation, heat removal and vessel geometry may interact adversely with other factors.[20] The dilute conditions typical in medicinal chemistry tend to minimise these potential issues so they are rarely significant at this stage. Catalytic reactions of course have many more factors to consider,[14,21] but at small scale early in a project's life, only the important factors need to be identified—optimisation can come later.

Table 10.2 Continuous and discrete factors that are likely to affect reactions.

	Continuous Factors	Discrete Factors
Typically important factors	Reagent stoichiometry	Reagent identity
	Base stoichiometry	Base identity
	Catalyst loading	Catalyst identity
	Addition rate (reagent/base)	Order of addition
	Concentration	Solvent identity
	Temperature	
	Time	
	pH	
Less commonly important factors	Agitation rate	Agitation method
	Pressure (gas)	Vessel geometry
	Additive stoichiometry	Additive identity
	Ligand stoichiometry	Ligand identity
	Metal:ligand ratio (pre-catalyst)	Catalyst addition method
	Surface area (solid reagents)	

The more difficult question to answer, and one that always arises, is how wide to set the ranges on the continuous factors. The general advice is that the range shouldn't be so wide as to be outside the normal scope for the reaction (where adverse results may be observed), but wide enough to observe and quantify any effect if it exists. This ultimately has to be a matter of judgement, but since chemists are skilled and knowledgeable practitioners of their science, this shouldn't be limiting. Common sense dictates some factors. For example, the pH range investigated for a bio-catalytic reaction will probably be in the narrow range tolerated by the enzyme; but reaction concentration may need to be varied widely before significant effects are observed.

As noted, typically only a few factors will be important for an initial investigation, which could give quick improvements in terms of yield or selectivity for the work involved. Therefore, a DoE of 8–16 experiments (plus replicates) is likely to cover the number of factors desired in a lightly fractionated and simple design. Four factors can be fully investigated at two levels in 16 (*i.e.* 2^4) experiments, or five factors could investigated in a half fractional DoE with almost no loss of information; but up to eight factors *could be investigated* in an DoE of 16 experiments, without losing the single factor effects or two factor interactions. Even relatively simple designs with low reaction numbers can generate worthwhile response surface models, as exemplified in two and three dimensional formats for two factors (Figure 10.3). More complex designs are also possible, but these are for increasing levels of optimisation, and are unlikely to be required by the medicinal chemist (see ref. 12 for some examples). However, development and manufacturing chemists would use them, and be able to build on even a simple design started in medicinal chemistry, so an initial DoE is always worth considering.

One problem that cannot easily be tackled with an initial DoE is how to deal with the discrete (*i.e.* qualitative) factors, such as solvent, catalyst, ligand and base. In more advanced studies, the technique of Principal Component Analysis (PCA) can be combined with DoE to turn these discontinuous discrete factors into continuous factors by breaking them down into measurable principal components, which can then be mapped by DoE. This has been especially applied to solvents,[23–25] but also to other reagents (*e.g.* ligands, organic bases, Lewis acids), for which Carlson has been a leading proponent.[26–28] Typically, a discrete factor is represented by three factors in a DoE to "map" it, which rapidly increases the number of factors that must be investigated. Consequently, this is likely to be well beyond the scope required by an early medicinal chemistry project although it is used on development projects, and lies behind some aspects of the work mentioned on solvents in Chapter 2.

On a last practical point, many chemists are deterred by the mathematics and equations involved in DoE. Whilst there is no doubt that an understanding of statistics will be helpful, there are a number of companies that produce software specifically for these applications (and offer training courses as well).[15] Furthermore, many large companies have a small group of

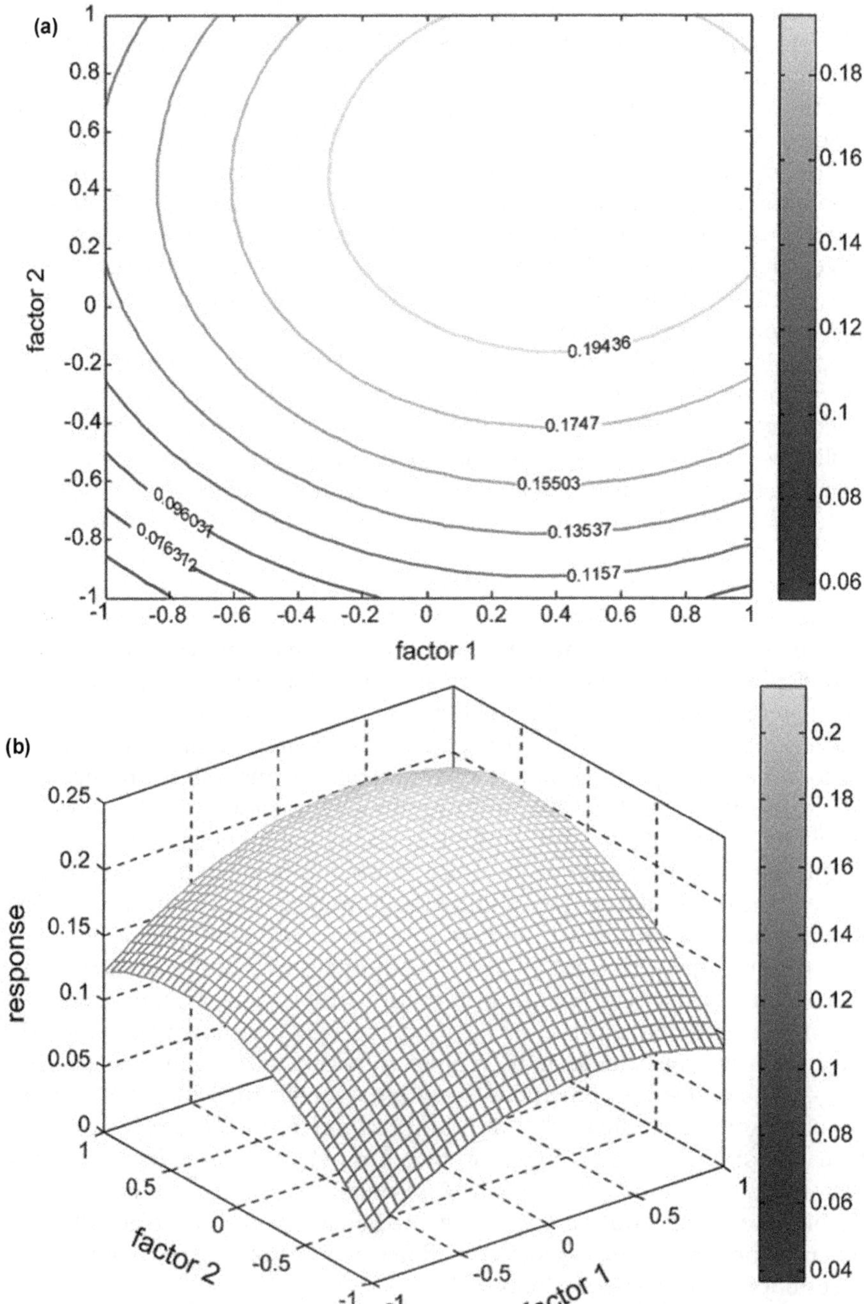

Figure 10.3 Example response surface model in two and three dimensional models for two factor interaction.
Reproduced with permission from Figure 12 in ref. 22: B. Dejaegher and Y. Vander Heyden, *J. Pharm. Biomed. Anal.*, 2011, **56**, 141, Elsevier.

statisticians who use their skills across multiple departments, and who are likely to be only too keen to offer their help on an unfamiliar problem and to support some practical science. Failing that, there is likely to be at least one mathematically literate chemist who is prepared to be trained up to be the local expert and to champion the cause of DoE—and this person will have the advantage of understanding the chemistry.

10.5 A Case Study

By way of final example, the study by Aggarwal and Owen[29] on the Heck reaction shown in Figure 10.4 illustrates a number of the foregoing points, and should be both relevant and appealing to medicinal chemists. The full details from this short publication are not re-iterated here, but in summary, the authors wanted to improve a reaction that gave ~40 : 60 ratio of cinnamaldehyde starting material to Heck product. They were keen to improve the yield and reduce the catalyst loading from 3 mol%, for which they thought that five factors might be important (base stoichiometry, catalyst loading, concentration, temperature and time), which as a full design would require 2^5 (*i.e.* 32) reactions. Ranges were set fairly wide in this case with justifications provided. As is common in DoEs, they decided to take multiple timed samples from each reaction performed, thereby increasing the data generated. Reactions were performed at 1 mmol scale in a parallel reactor, thus reducing experimental variation, and analysis was performed by GC-MS. Knowing that higher order interactions were unlikely to be significant, they decided to run a half factorial design of 16 reactions, with four replicates as the centre points, which gave a manageable 20 reactions in total.

From the initial DoE, they found that temperature was the most important factor for this reaction, with catalyst loading and concentration having lesser effects, and a similar effect for the interaction of temperature and concentration. Base stoichiometry was found to have almost no effect at all (within the ranges studied). Overall, the results indicated that lower catalyst loading could probably be achieved at higher dilutions and temperatures than they had so far investigated. They then conducted a second iteration using a response surface model design to optimise the conditions further (for which the actual responses for volume and catalyst loading looked very similar to Figure 10.3). Catalyst loading, concentration, temperature and time were

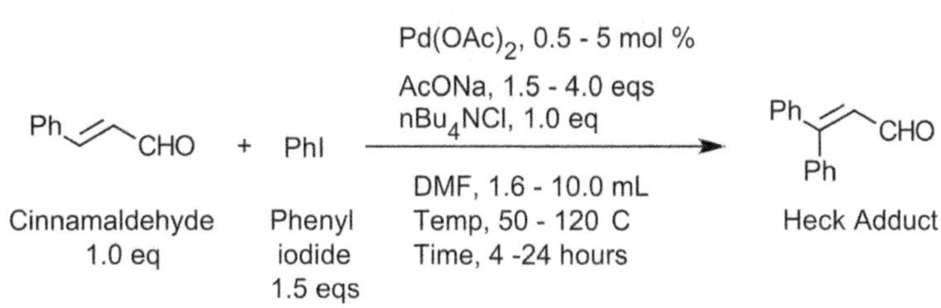

Figure 10.4 Heck reaction optimised by DoE (with widest ranges investigated shown).

studied, with the base stoichiometry now fixed. In conclusion, they found that by performing the reaction at 120 °C instead of 60 °C, and increasing the dilution 2.5 times, only 0.5 mol% catalyst was needed to give a reproducible GC-MS *conversion* of 98%, and an *isolated yield* of 89% (the previous conditions had given 58% conversion with an isolated yield of 40% in their hands, or 59% claimed in the previous best literature case). This study illustrates a number of points:

- The first iteration DoE identified the key factors and gave significant *immediate* improvements
- It also indicated that much better results might be obtained outside the design space investigated
- It showed the "direction of travel" required to reach this space
- The half factorial proved entirely adequate, the full factorial was not required
- One second order interaction was identified (concentration and temperature)
- A response surface model was built on the initial results for the full optimisation study
- The use of an automated reactor and solution GC-MS analysis aided rapid project delivery

What is particularly notable about this reaction was that it had been studied before by OVAT type techniques and further improvement had been thought impossible. Lastly, this DoE study was conducted in only 3 weeks.

10.6 Summary

In conclusion, this chapter has sought to commend DoE to medicinal chemists for use at the appropriate time and place on their projects. Where DoE can be used within the admittedly tight time constraints of medicinal chemistry, it will often:

- Benefit the project immediately in terms of improved reaction understanding, material throughput, yield, quality and simplified work-ups
- Indicate where further improvements can be found
- Provide an excellent foundation for scale-up colleagues to build upon
- Save time overall, both within the medicinal chemistry and process development phases of a project
- Lead to greener processes

Acknowledgements

The author thanks Helen Sneddon (GSK, Stevenage) for helpful initial discussions on this chapter.

References

1. G. E. P. Box, W. G. Hunter and J. S. Hunter, *Statistics for Experimenters – An Introduction to Design, Data Analysis, and Model Building*, Wiley, New York, 1978.
2. *Design and Analysis in Chemical Research*, ed. R. L. Tranter, CRC Press, Sheffield, UK, 2000.
3. R. Carlson and J. E. Carlson, *Design and Optimization in Organic Synthesis*, Elsevier, Amsterdam, 2nd edn, 2005.
4. M. R. Owen, C. Luscombe, L.-W. Lai, S. Godbert, D. L. Crookes and D. Emiabata-Smith, *Org. Process Res. Dev.*, 2001, **5**, 308.
5. J. T. Kuethe, D. M. Tellers, S. A. Weissman and N. Yasuda, *Org. Process Res. Dev.*, 2009, **13**, 471.
6. S. Sjövall, L. Hansen and B. Granquist, *Org. Process Res. Dev.*, 2004, **8**, 802.
7. L. S. Riter, O. Vitek, K. M. Gooding, B. D. Hodge and R. K. Julian, *J. Mass Spectrom*, 2005, **40**, 565.
8. G. Franceschini and S. Macchietto, *Chem. Eng. Sci.*, 2008, **63**, 4846.
9. C.-F. Mandenius and A. Brundin, *Biotechnol. Prog.*, 2008, **24**, 1191.
10. C. Pierlot, L. Pawlowski, M. Bigan and P. Chagnon, *Surf. Coat. Technol.*, 2008, **202**, 4483.
11. V. A. Sakkas, M. A. Islam, C. Stalikas and T. A. Albanis, *J. Hazard. Mater.*, 2010, **175**, 33.
12. H. Tye, *Drug Discov. Today*, 2004, **9**, 485.
13. D. Lendrem, M. Owen and S. Godbert, *Org. Process Res. Dev.*, 2001, **5**, 324.
14. J. D. Moseley and P. M. Murray, *J. Chem. Technol. Biotechnol.*, 2014, **89**, 623.
15. (a) www.umetrics.com; (b) www.statease.com; (c) www.jmp.com; (d) www.minitab.com.
16. But not always! For example, Denmark ran a fractionated DoE of ~175 reactions from a total of 1440 possible reactions covering multiple factors in a complex catalytic reaction, from which good results were obtained. Impressive though this was, this is hardly appealing to the time-pressured medicinal chemist, even with the assistance of automation. See: S. E. Denmark and C. R. Butler, *J. Am. Chem. Soc.*, 2008, **130**, 3690.
17. For example, in the author's own experience, a DoE on a low yielding (17%) aromatic bromination based on a medicinal chemistry route revealed that the reaction could not be improved; however, the alternatives were worse. See: J. D. Moseley, W. O. Moss, M. J. Welham, C. L. Ancell, J. Banister, S. A. Bowden, G. Norton and M. J. Young, *Org. Process Res. Dev.*, 2003, **7**, 58.
18. A. Gopalsamy, M. Shi and R. Nilakantan, *Org. Process Res. Dev.*, 2007, **11**, 450.
19. J. P. Larkin, C. Wehrey, P. Boffelli, H. Lagraulet, G. Lemaitre, A. Nedelec and D. Prat, *Org. Process Res. Dev.*, 2002, **6**, 20.

20. Although for a rare example of these effects at small scale see: A. J. J. Lennox and G. C. Lloyd-Jones, *J. Am. Chem. Soc.*, 2012, **134**, 7431.
21. J. D. Moseley, P. M. Murray and S. N. G. Tyler, *Org. Process Res. Dev.*, 2013, **17**, 40.
22. B. Dejaegher and Y. Vander Heyden, *J. Pharm. Biomed. Anal.*, 2011, **56**, 141.
23. A. R. Katritzky, D. C. Fara, M. Kuanar, E. Hur and M. Karelson, *J. Phys. Chem. A*, 2005, **109**, 10323.
24. M. Allesøa, J. Rantanena, J. Aaltonenb, C. Cornetta and F. van den Berg, *J. Chemometrics*, 2008, **22**, 621.
25. R. A. Stairs and E. Buncel, *Can. J. Chem.*, 2011, **84**, 1580.
26. R. Carlson, J. Carlson and A. Grennberg, *J. Chemometrics*, 2001, **15**, 455.
27. R. Carlson and J. E. Carlson, *Org. Process Res. Dev.*, 2005, **9**, 680.
28. R. Carlson, G. Simonsen, A. Descomps and J. E. Carlson, *Org. Process Res. Dev.*, 2012, **16**, 1371.
29. V. K. Aggarwal, A. C. Staubitz and M. Owen, *Org. Process Res. Dev.*, 2006, **10**, 64.

CHAPTER 11

Pd-catalysed Cross-couplings for the Pharmaceutical Sector and a Move to Cutting-edge C–H Bond Functionalization: Is Palladium Simply Too Precious?

IAN J. S. FAIRLAMB

University of York, Department of Chemistry, Heslington, York YO10 5DD, UK
Email: ian.fairlamb@york.ac.uk

Cross-coupling reactions are widely used in the pharmaceutical sector to access structurally diverse products containing arylated and heteroarylated products.[1] C–C, C–N and C–O bond-forming processes are firmly embedded in the pharmaceutical scientist's armoury of synthetic tools. Some of these ubiquitous reactions are collated in Scheme 11.1. Their importance cannot be over-stated–they are critical transformations for Research and Development (R&D), with many reactions appearing downstream in Process and Manufacturing (P&M).[2]

The catalyst of choice for many of these transformations is the precious metal Pd, which is seen as a *critical metal* going forwards (present at very low

RSC Green Chemistry No. 46
Green and Sustainable Medicinal Chemistry: Methods, Tools and Strategies for the 21st Century Pharmaceutical Industry
Edited by Louise Summerton, Helen F. Sneddon, Leonie C. Jones and James H. Clark

Published by the Royal Society of Chemistry, www.rsc.org

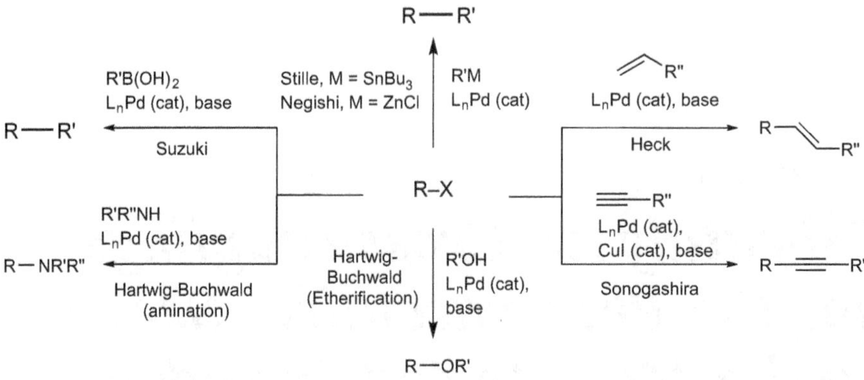

Scheme 11.1 Common cross-coupling reactions used in the pharmaceutical sector.

concentrations in the earth's crust <1 μg kg^{-1}). Reactions are not limited to Pd however, with Mn, Co, Fe, Ni and Cu seeing considerable attention in the last 5–10 years within the literature.[3] Whilst the advantage of element abundance is obvious for the latter metals (referred to as base metals), current usage levels also need to be seriously considered. For example, current usage of Cu could see it being designated as a rare metal within 60 years. On the other hand, Pd will be exhausted within 300 years at current usage levels. This point is often over-looked in making the case for Cu over Pd. The argument for using base metals, such as Fe, is clearer. Fe chemistry is particularly fascinating, but when compared reaction-by-reaction to Pd, it is way behind the myriad of transformations that are possible using the precious metal.

This is undoubtedly an exciting time for cross-coupling chemistry, particularly the opportunities that lie ahead with base metals, *but* it is important to acknowledge that the field of Pd chemistry will continue to play a key role in synthetic chemistry, with the prediction of prominence for Pd-catalysed processes for at least the next 25 or so years, perhaps longer. To suggest that base metals will see the fading-out of precious Pd in industrial processes over the coming years is both naive and unlikely. The primary reason is that cross-coupling by Pd currently supersedes most other metals, particularly in terms of substrate scope, process versatility and catalyst loading. Moreover, there are ways of utilising Pd that can be in keeping with the principles of Green Chemistry.[4]

Is there a case against Pd? The cost of Pd has continued to increase in recent years, however the value of Pd has traditionally been highly volatile. In 2001, the cost per ounce was ~$1100, whereas in 2009 it was *ca.* $200, and in March 2014 it is ~$770 (with some price stability seen between 2013/2014). A more challenging issue for Pd is its acute toxicity—it is a metal of serious safety concern, according to the European Medicines Agency (permitted daily exposure if taken orally should be <100 μg day^{-1} or 10 ppm; if inhaled the Pd content needs to be <10 μg day^{-1} or 1 ppm).[5] The pharmaceutical sector has thus focussed much attention in terms of ensuring that final drug products have minimal Pd contamination (<1 ppm levels). Drug compounds with their dense functionalization may by their very nature contain palladaphilic functional groups and/or heteroaromatic motifs. Harder nitrogen donors in particular bind with oxidised Pd(II). The issue of Pd impurities is a particular challenge on the smaller scale, especially if one considers that 5 mol% Pd is common in synthetic research laboratories, whereas significantly less might be used in a process laboratory. Nevertheless, it is often necessary to employ Pd scavengers (*e.g.* thiolated silicas) to remove trace Pd post-formation of the final drug product.

It is perhaps surprising to not see more processes being developed, which can be more efficient when run at lower Pd catalyst loadings. For example, the optimal catalyst loading for a standard Heck cross-coupling of an aryl halide with a terminal alkene is 0.01 mol% (*ca.* 37 ppm Pd).[6,7] Higher Pd catalyst loadings lead to ineffective catalysis. A similar observation was recorded in Sonogashira cross-coupling reactions using palladacycles as a precatalyst.[8] These observations are rarely exploited, even though Leadbeater showed that for certain cross-couplings sub-ppm levels of Pd are enough to catalyse their reaction at higher temperatures.[9]

Given that trace Pd can mediate cross-coupling processes, it is important that in the development of base-metal catalysed cross-couplings the Pd concentration (down to ppb levels) is determined and appropriate controls conducted. Bedford showed that Pd contaminants were catalytically active in a supposedly Fe-catalyzed Suzuki–Miyaura cross-coupling methodology, employing common aryl bromide substrates.[10]

A late-stage Pd-mediated transformation could be seen as a serious problem, where Pd contamination is concerned. Arguably, there is good reason to 'box-clever' with Pd. Recent developments in catalyst design can help, but with so many catalysts (mostly precatalysts) being available, where does the industry focus their screening efforts? Mechanistic insight can help considerably. The widely used Suzuki–Miyaura cross-coupling (SMCC) reaction is a good example to mention. The mechanism is much more complicated than the 'textbook mechanism' would have you believe. For example, for over 30 years it has generally been accepted that the oxidative addition product reacted directly with an activated organoboronate species (*e.g.* formed by reaction of organoboronic acid with base) in the transmetallation step. However, the formation of a 'Pd–OH' critical species is required to activate the neutral organoboronic acids.[11] Further details show that the concentration of base, *i.e.* trickle-feed of low concentrations of hydroxide (and associated counter-cation) are central to the success of many SMCCs.

Other studies have also demonstrated that lower Pd catalyst loadings can in some cases lead to more efficient substrate turnover.[6] This counter-intuitive observation is the result of Pd-agglomeration leading to inactive higher-order Pd phases under certain reaction conditions. On the contrary, under some reaction conditions Pd nanoparticles (PdNPs) are active catalyst species. Indeed, a heterogeneous SMCC has been reported, which necessitated the use of Synchrotron radiation to probe the structure of the Pd catalyst phase under specific working conditions.[12] In keeping with this finding, many tailored PdNPs have been developed for use in SMCCs and Heck alkenylation particularly.[13] Currently, catalyst performance is not yet commensurate with high-performance homogeneous catalysts. Pd leaching is still an issue, even with well-defined and encapsulated PdNPs, including commercially useful catalysts (*e.g.* ENCAT™).[14] In many circumstances, particularly in P&M, Pd/C catalysts are the preferred 'heterogeneous' catalyst, which is known to leach single atom/small Pd cluster species into solution. This latter comment highlights that a large number of cross-coupling reactions are influenced by the presence of both homogeneous and heterogeneous species, the latter acting as reservoirs of 'active Pd' or being active in their own right on the catalyst surface. One major advantage of a truly heterogeneous cross-coupling at Pd is that Pd contamination can be minimised. Moreover, Pd catalyst recycling is possible, and this has been fully realised in the pharmaceutical sector over the last 2–3 years. In addition, it should be mentioned that processes in flow devices become viable.

The case to 'box clever'? Over the last decade, researchers have focussed on improving cross-coupling processes by developing remarkable selective C–H bond functionalization strategies. C–C, C–O and C–X bond formation is now feasible in a raft of examples. The clear advantage is that the need for substrate prefunctionalization is removed. The reaction of an organohalide (RX) with a Grignard reagent (R′MgX, represented by R′M) to give R–R′ and MgX_2 (side product) is a classical reaction where both starting materials are prefunctionalized. Replacement of R′MgX with a nucleophilic R′H substrate (together with a base) can give the same R–R′ cross-coupled product and HX side product. The idealised atom efficient transformation, based on substrates at least, is the reaction of RH and R′H to give R–R′ (and formally H_2 as a side product). In such a case, an oxidant is required, for example air, to give H_2O as the side product, the formation of which helps drive the reaction on a thermodynamic basis. Generalized examples are collated in Scheme 11.2, along with a simplified mechanism to describe the concepts underlying the different approaches.

Various processes have been developed and many mechanistic pathways [including all oxidation states between Pd(0) and Pd(IV)] are energetically feasible, making the multitude of possible transformations truly remarkable. The frontier-leading reactions involve formal loss of H_2. Processes that can be considered as being of intermediate efficiency involve formation of HX (base needed) or MH (oxidant needed), producing MX by-products (a downstream waste issue). The case for these processes has been made elsewhere.[3] It is important to note that additives facilitating the selective C–H functionalization are oftentimes numerous—it is common to hear the phrase 'black-box'. Moreover, in some cases stoichiometric co-oxidants are absolutely essential [*e.g.* Ag(I) or Cu(II)], which on a small scale (<10 g) is not an issue, but on a larger scale (~kg) can become so.

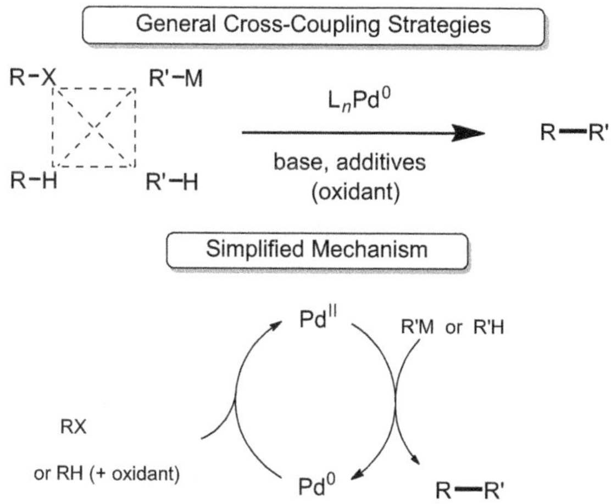

Scheme 11.2 A generic cross-coupling process, which shows how C–H bond functionalization reactions and classical cross-couplings compare, in terms of their substrates.

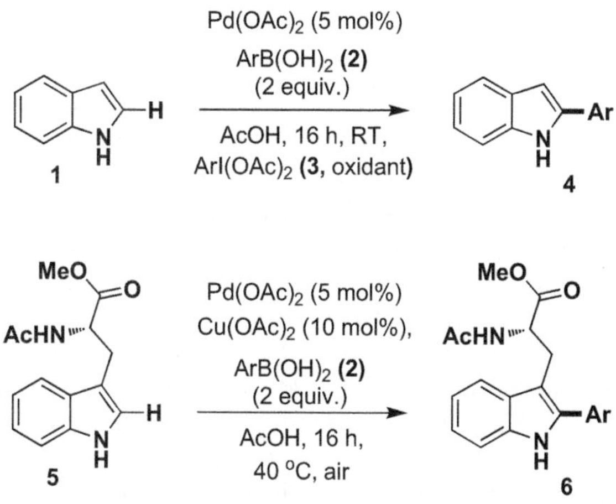

Scheme 11.3 Oxidative cross-coupling reactions involving indoles and organoboronic acids: (a) an exemplar indole arylation reaction; (b) arylation of a more complicated tryptophan derivative.

These criticisms aside, oxidative processes have emerged in spectacular fashion since 2007. Whilst teaching undergraduates, it would be inconceivable to suggest that two nucleophilic coupling partners can be reacted together. It sounds like an extreme case, but taking two organometallic species and reacting them together in the presence of an oxidant works, and offers an alternative strategy in some organic systems. More viable, and perhaps more chemically intuitive, is to react an aryl or heteroaryl coupling partner with an organoboronic acid under oxidative conditions. An example is given in Scheme 11.3 for indole and derivatives, which are heteroaromatic motifs of particularly therapeutic interest. For example, the methodology published by Sanford and co-workers[15] is particularly attractive (**1**→**4**), which employs an aryliodonium(III) diacetate (**3**) species as the co-oxidant (in this chemistry both the aryl from arylboronic acid (**2**) and aryliodonium diactate species can in principle appear in the cross-coupled product). Larrosa[16] and Zhang[17] have reported complementary methodologies, the latter involving a Cu(II) catalyst and air as the oxidant system. It is of particular note that Zhang's conditions could be applied to more complex substrates, such as a protected tryptophan derivatives (**5**→**6**).[18] In such a case it was noted that the alternative arylation in the presence of phenyliodonium(III) diacetate and phenylboronic acid also worked well, where Pd nanoparticles (PdNPs) were found to be relevant catalyst species.

Fagnou and co-workers[19] addressed the issue of regioselectivity in indole functionalization in a paper that was perhaps one of the most important reported in the C–H bond activation field between 2000 and 2010. It was found that a double C–H bond functionalization reaction was possible utilising Pd(II) and appropriate stoichiometric oxidants. In the presence of $Cu(OAc)_2$ the arylated C3-regioisomer was formed as the major regioisomer (**7**→**9**), whereas using AgOAc the arylated C2 regioisomer was preferred (**7**→**8**). The mechanistic dichotomy is difficult to explain, although it is tempting to suggest the involvement of mixed Pd–Ag and Pd–Cu species, which would go some way to explaining the divergence in regioselectivity (note: other variables could also explain the difference recorded) (Scheme 11.4).

Whilst the leading nature of Fagnou's methodology is clear, from a waste perspective there are again stoichiometric waste metal side products, presumably in the form of Cu(I) and Ag(0), which need to be dealt with, which could be an issue on scale-up.

In terms of predicting the reactivity of C–H bonds in specific (hetero)arene substrates, it is worth consulting the review written by Gorelsky.[20] Here, the origin of the regioselectivity in many Pd-catalysed C–H bond functionalization reactions (involving metalation–deprotonation processes) has been examined, leading to the proposal that the C–H bond acidities are the principle factor in determining the position of functionalization, especially in electron-deficient (hetero)arenes. He goes on to explore the contributions of distortion and electronic interactions of the Pd intermediate species (*e.g.* the interaction of the substrate and therefore ligand with the Pd centre). A metalation–deprotonation step occurs in a concerted manner, which is

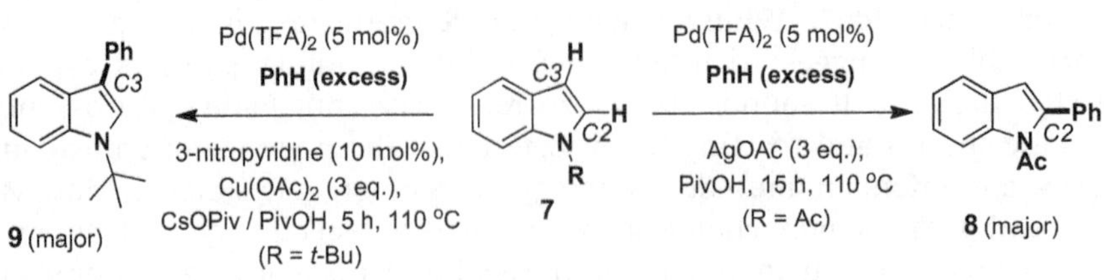

Scheme 11.4 Fagnou's double C–H bond functionalization process.

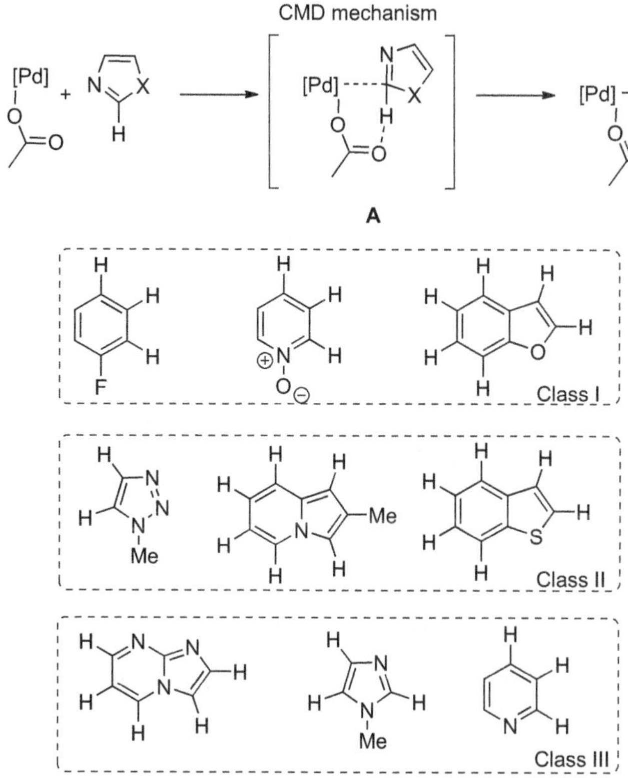

Scheme 11.5 The CMD mechanism and classification of different (hetero)arene substrates. The CMD mechanism is shown for one step in an unspecified catalytic cycle. Typically, the Pd(II) centre contains another aryl group, which would see reductive elimination from intermediate **B** to give the organic product, leading to regeneration of the active Pd catalyst (not shown).

commonly referred to as a Concerted Metalation Deprotonation or a CMD process (Scheme 11.5), involving transition state structure **A** (which converts to intermediate **B**).[21,22]

Gorelsky divided both arenes and heteroarenes into three classes, taking into account the (hetero)arene distortion energy differences (class I), the interaction energies with the metal catalyst as a reactivity determining factor (class II), and where distortion and interaction energies enhance the functionalization of one particular C–H bond (class III). In most class I and III substrates the regioselectivity is derived from C–H bond acidity, that is functionalization (arylation) occurs at the most acidic C–H bond. Perutz and Eisenstein[23] have shown that the Pd–C bond strength is also a critical factor in determining the preferred regioselectivity, although Gorelsky makes the case that the CMD activation barriers do not show a correlation with the C–H bond acidities and metal-(hetero)arene bond energies.

Whilst there are relatively few leading examples of C–H functionalization reactions reported from within the pharmaceutical sector, a few key studies have emerged. For example, at Pfizer a high-yielding and scalable C–H bond functionalization process[24] (involving loss of HCl) was developed to facilitate kilogram quantities of oxindole **11** to be prepared; oxindole **11** is a key intermediate *en route* to a serine palmitoyl transferase inhibitor **12** (Scheme 11.6).[25] The Pd-catalyzed reaction of α-chloroacetanilide **10** in the presence of Et_3N, $Pd(OAc)_2$ and 2-(di-*tert*-butylphosphino)biphenyl ligand afforded oxindole **13** in 70–80% yields. An optimised solvent combination of 2-MeTHF/2-propanol (4 : 1, v/v) was found to be key to the success of the reaction. On scale-up, the reaction of 10.95 kg of α-chloroacetanilide **10** gave 7.7 kg of oxindole product **11** (76% yield). Of particular note was the finding that the preferred ligand from the original Buchwald study, 2-(di-*tert*-butylphosphino)biphenyl, could be substituted for Pfizer's non-proprietary and commercially available ligand, 5-(di-*tert*-butylphosphino)-1′,3′,5′-triphenyl-1′*H*-[1,4′]-bipyrazole,[26] also known as BippyPhos **13**. On 25 g scale, an 84% yield of oxindole **11** was obtained using the BippyPhos ligand. The new route eliminated the need for NaH and two chromatographic purifications in the original route. It was noted that Pd impurities were present in the product at a level of 100–800 ppm, although no attempts were made to remove this residual Pd in this particular process. It was further acknowledged that it is necessary to lower the Pd catalyst loading and ligand to ensure that the process becomes cost-effective.

Lilly Research Laboratories described[27] a remarkable intermolecular benzylation reaction involving C–H bond functionalization of an imidazopyridazine **14** *en route* to the synthesis of LY2784544, which is a JAK2-V617F inhibitor. The reaction is regioselective for the C3-position, and from an extensive ligand screen PPh_3 came out as the ligand of choice, to give **16** in good yield. K_2CO_3 was found as the optimal base and 1,4-dioxane as the preferred solvent. The 1,4-dioxane (a suspect carcinogen[28]) could be

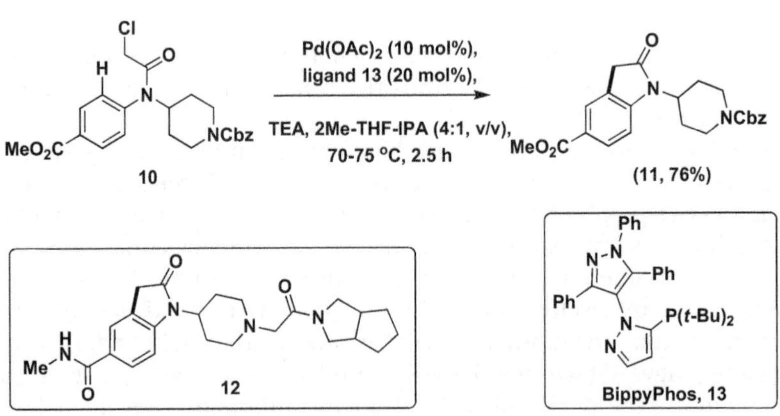

Scheme 11.6 Pfizer's oxindole formation by a selective C–H bond functionalization reaction.

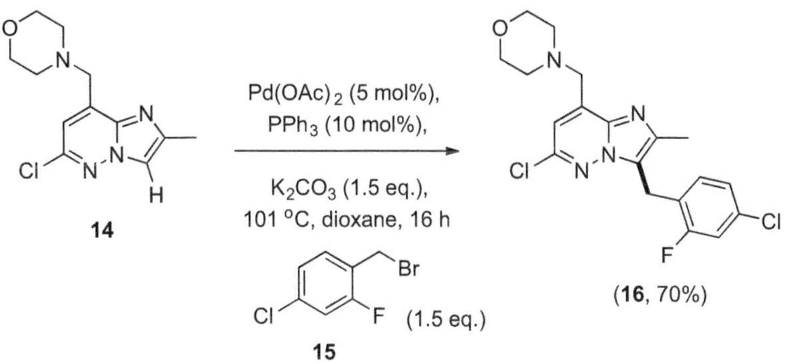

Scheme 11.7 Benzylation of imidazopyridazine **14**, to give **16**, by a selective C–H bond functionalization reaction.

substituted for anisole, although this led to a slight drop in yield from 70% to 61%. More generally speaking it is clear that more greener and cleaner solvents are needed for C–H bond functionalization reactions, especially as solvent is often the major contributor to sustainability metrics (see also Chapter 4, Renewable Solvent Selection in Medicinal Chemistry) (Scheme 11.7).

Green metrics (*e.g.* atom economy and step economy) are seriously considered when developing new synthetic transformations for the pharmaceutical sector (*i.e.* in process and manufacturing). Historically, researcher development teams have focussed less on these considerations, leaving process scientists to solve atom and step economic issues, in addition to practical issues that arise on scale-up. Critically speaking, the landscape has changed in recent years and green considerations are increasingly taken into account in the drug design process at many pharma-based companies.[2]

It is worthy of note that Pd nanoparticles have been successfully used as catalysts for the C–H bond functionalization of various heteroaromatic,[18,29] highlighting that the reactions are not limited to supposedly homogeneous Pd catalyst sources.

Finally, it is worth highlighting that it is not always necessary to turn to a Pd-catalyzed reaction for a given synthetic target. For example, a critical comparison of a Pd-catalysed cross-coupling approach (Hartwig–Buchwald reaction) with a classical S_NAr reaction for the amination of certain heteroaromatic chlorides revealed that the latter approach was highly attractive.[30]

Summary

- Palladium is a critical metal for cross-coupling processes, despite being precious, and will continue to be so for years to come.
- It is possible to 'box-clever' with Pd, either through the use of immobilized Pd catalysts, including stabilized nanoparticles, or ultra-low catalyst loadings (low ppm levels). The latter has not been fully

exploited, despite several studies showing that lower Pd catalyst loadings can lead to more effective cross-coupling catalysis.

- C–H bond functionalization chemistry is now a mature research field. Many (hetero)arenes can be functionalized (*i.e.* arylated) in a predictable manner.
- Moving forward, it will be important to develop more environmentally benign solvents (beyond common polar aprotic solvents) for C–H bond functionalization chemistry.

Financial Disclosure

The author has previously received EPSRC funding in support of some of the original publications cited in this article. He has current funding from: (1) Bayer Cropscience (Frankfurt, Germany), which supports a PhD studentship (2012–2016); (2) Innovative Medicines Initiative 4 (grant agreement no 115360), Chem 21 project (including several pharmaceutical companies within the consortium), which directly supports two PhD studentships (2012–2015) and 1 post-doctoral scientist. The research in each project relates to the subject matter of this specific review chapter. The author has also received funding from Bruker Ltd., which supports a PhD studentship (2012–2016) investigating new NMR technologies for studying Pd catalysed processes and ligand exchange.

References

1. J. S. Carey, D. Laffan, C. Thomson and M. T. Williams, *Org. Biomol. Chem.*, 2006, **4**, 2337–2347.
2. M. C. Bryan, B. Dillon, L. G. Hamann, G. J. Hughes, M. E. Kopach, E. A. Peterson, M. Pourashraf, I. Raheem, P. Richardson, D. Richter and H. F. Sneddon, *J. Med. Chem.*, 2013, **56**, 6007–6021, and references cited therein.
3. Aspects of base metal catalysis relating to cross-coupling processes are covered generally in the following book: C-H and C-X Bond Functionalization. Transition Metal Mediation. RSC Cata*lysis*, ed. X. Ribas, 2013, No. 11.
4. P. T. Anastas and J. C. Warner, *Green Chemistry: Theory and Practice*, Oxford University Press, New York, 1998, p. 30.
5. European Medicines Agency, Pre-authorisation Evaluation of Medicines for Human Use, 'Guideline on the specification limits for residues of metal catalysts' pp. 4–13, (2007). Interestingly, the human average dietary intake of Pd is estimated to be $\sim$2 μg/day.
6. A. H. M de Vries, J. M. C. A. Mulders, J. H. M. Mommers, H. J. W. Henderickx and J. G. de Vries, *Org. Lett.*, 2003, **5**, 3285–3288.
7. M. Reetz and J. G. de Vries, *Chem. Commun.*, 2004, 1559–1563.
8. I. J. S. Fairlamb, A. R. Kapdi, A. F. Lee, G. Sanchez, G. Lopez, J. L. Serrano, L. Garcia, J. Perez and E. Perez, *Dalton Trans.*, 2004, 3970–3981.

9. R. K. Arvela, N. E. Leadbeater, M. S. Sangi, V. A. Williams, P. Granados and R. D. Singer, *J. Org. Chem.*, 2005, **70**, 161–168.
10. R. B. Bedford, N. Nakamura, N. J. Gower, M. F. Haddow, M. A. Hall, M. Huwe, T. Hashimoto and R. A. Okopie, *Tetrahedron Lett.*, 2009, **50**, 6110–6111.
11. A. J. J. Lennox and G. C. Lloyd-Jones, *Angew. Chem., Int. Ed.*, 2013, **52**, 7362–7370.
12. P. J. Ellis, I. J. S. Fairlamb, S. F. J. Hackett, K. Wilson and A. F. Lee, *Angew. Chem., Int. Ed.*, 2010, **122**, 1864–1868.
13. C. Deraedt and D. Astruc, *Acc. Chem. Res.*, 2014, **47**, 494–503.
14. S. J. Broadwater and D. T. McQuade, *J. Org. Chem.*, 2006, **71**, 2131–2134.
15. N. R. Deprez, D. Kalyani, A. Krause and M. S. Sanford, *J. Am. Chem. Soc.*, 2006, **128**, 4972–4973.
16. N. Lebrasseur and I. Larrosa, *J. Am. Chem. Soc.*, 2008, **130**, 2926–2927.
17. Z. Liang, B. Yao and Y. Zhang, *Org. Lett.*, 2010, **12**, 3185–3187.
18. (a) T. J. Williams, A. J. Reay, A. C. Whitwood and I. J. S. Fairlamb, *Chem. Commun.*, 2014, **50**, 3052–3054; (b) A. J. Reay, T. J. Williams and I. J. S. Fairlamb, *Org. Biomol. Chem.*, 2015, **13**, 8298.
19. D. R. Stuart and K. Fagnou, *Science*, 2007, **316**, 1172–1175.
20. S. I. Gorelsky, *Coord. Chem. Rev.*, 2013, **257**, 153–164.
21. The alternative mechanism involves a more classical electrophilic aromatic substitution (S_EAr) process, involving a Wheland-type intermediate.
22. This process is also referred to as ambiphilic metal–ligand activation (AMLA-6), where the 6 refers to a six-membered transition state. AMLA-6 and CMD, in essence, describe the same mechanism, see: D. L. Davies, S. M. A. Donald and S. A. Macgregor, *J. Am. Chem. Soc.*, 2005, **127**, 13754–13755.
23. J. Guihaume, E. Clot, O. Eisenstein and R. N. Perutz, *Dalton Trans.*, 2010, **39**, 10510–10519.
24. E. J. Hennessy and S. L. Buchwald, *J. Am. Chem. Soc.*, 2003, **125**, 12084–12085.
25. A. W. Kiser, J. Magano, R. J. Shine and M. H. Chen, *Org. Process Res. Dev.*, 2012, **16**, 255–259.
26. G. J. Withbroe, R. A. Singer and J. E. Sieser, *Org. Process Res. Dev.*, 2008, **12**, 480–489.
27. A. N. Campbell, K. P. Cole, J. R. Martinelli, S. A. May, D. Mitchell and P. M. Pollock, *Org. Process Res. Dev.*, 2013, **17**, 273–281.
28. The carcenogenic properties of 1,4-dioxane are discussed in: K. S. Surprenant, *Ullmann's Encyclopedia of Industrial Chemicals*. Wiley-VCH Verlag, 2002, pp. 309–314.
29. C. G. Baumann, S. De Ornellas, I. J. S. Fairlamb, J. P. Reeds, T. E. Storr and T. J. Williams, *Tetrahedron*, 2014, **70**, 6174–6187.
30. K. Walsh, H. F. Sneddon and C. J. Moody, *ChemSusChem*, 2013, **6**, 1455–1460.

CHAPTER 12

The Growing Impact of Continuous Flow Methods on the Twelve Principles of Green Chemistry

A. JOHN BLACKER,* JESSICA R. BREEN, RICHARD A. BOURNE AND CHRISTOPHER A. HONE

Institute of Process Research and Development, School of Chemistry and School of Chemical and Process Engineering, University of Leeds, Leeds LS2 9JT, UK
*Email: j.blacker@leeds.ac.uk

12.1 Introduction

The last decade has seen remarkable growth in the development of continuous flow reactions. Concurrently, the area of Green Chemistry has developed rapidly in response to the necessity to reduce the environmental impact of chemicals. This chapter considers the beneficial combination, and potential impact, of continuous flow methodology with the twelve Principles of Green Chemistry as developed by Anastas.[1] Whilst the application of continuous flow reactions is changing the way medicinal chemists are making and trying to identify lead bioactive compounds, the main impact of flow methods on green chemistry is in process development and manufacture, where the volumes of materials produced are far greater.

RSC Green Chemistry No. 46
Green and Sustainable Medicinal Chemistry: Methods, Tools and Strategies for the 21st Century Pharmaceutical Industry
Edited by Louise Summerton, Helen F. Sneddon, Leonie C. Jones and James H. Clark

Published by the Royal Society of Chemistry, www.rsc.org

The chapter is organised by discussion of each principle in turn with illustrative examples of published flow methods.

12.2 Principle 1: Prevent Waste Instead of Treating it

This principle is one of the most powerful. It aligns well with the flow process manufacturing aims of good economics, high efficiency and high throughput. The emphasis is on good process design from the outset, where waste can be prevented by designing both chemical transformation and product isolation procedures to avoid generating by-products, solvent waste, reagents or additive waste, impurities, catalyst waste and net energy input or output.

The least wasteful strategy for synthesizing a product is shown in eqn (12.1), in which no by-product is formed and the reaction is 100% atom-efficient.

$$A + B \rightarrow C \tag{12.1}$$

Reactions of this type include: addition to unsaturated substrates, pericyclic reactions and rearrangements.[2] Most reactions of this type require high-energy conditions or a catalyst. This makes this type of reaction more suitable to flow than batch. The high surface area to volume that tubular flow systems provide enables access to short-duration high-energy conditions (thermal, microwave, photochemical, electrochemical), which are suitable for making complex organic chemicals. An example of this is the DuPont nickel catalyzed, continuous gas–liquid phase addition of HCN to butadiene to make the nylon intermediate adiponitrile with low waste levels.[3,4]

If reactants A and B, and product C, are fluids, then continuous flow, solvent-free conditions may be suitable. This does, however, pose practical problems of efficient mixing between A and B, which can be made more difficult if either is viscous, or the reactions are fast. Within the polymer and food industries there are examples of both continuous solid and melt extrusion processes, though few involve reactive chemistry. If the reaction is selective then product C can be isolated directly, otherwise its higher molecular weight may make it difficult to separate without using a solvent, negating the aim, unless it is a solid. An example of a solvent-free continuous flow process is the lipase-catalyzed esterification of fatty esters using alcohols, as shown in Scheme 12.1.[5]

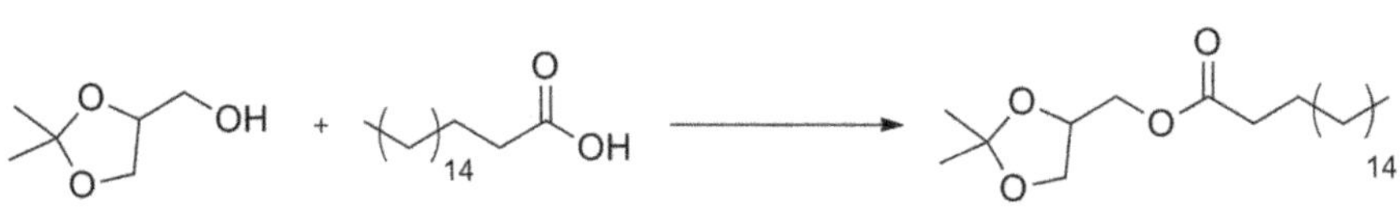

Scheme 12.1 Lipase-catalyzed esterification of fatty esters under solvent-free conditions.

12.3 Principle 2: Design Atom-efficient Methods

The most atom-efficient strategy is that in eqn (12.1), however most reactions fall into the type shown in eqn (12.2). Therefore, it is instructive to discuss this with respect to a continuous flow environment.

$$A + B \rightarrow C + D \qquad (12.2)$$

The molar ratio of product C, or by-product D, to reactants A and B, defines the atom efficiency. The most efficient reactions are those in which D is a small molecule, such as CO_2, CO, H_2O, NH_3, NaCl or $NaHSO_4$. Moreover, these are most easily handled in a flow environment and might be separated from the product either directly or through continuous washing.

Accordingly, reactants A and B are best designed with this in mind, noting that the way in which these are made is also important. For example, flow methodology allows both safe handling and control of selectivity for the use of gaseous F_2 (50% atom efficient) in preference to diethylaminosulfur trifluoride (12% atom efficient) for electrophilic fluorination of suitable organics (Scheme 12.2).[6]

Similarly, it is possible to make an epoxide from an alkene in flow using H_2O_2 with an oxometallate catalyst (47% atom efficient) and is much preferred to *m*CPBA epoxidation (9% atom efficient) in which by-product separation is by aqueous extraction which all becomes waste (Scheme 12.3).[7]

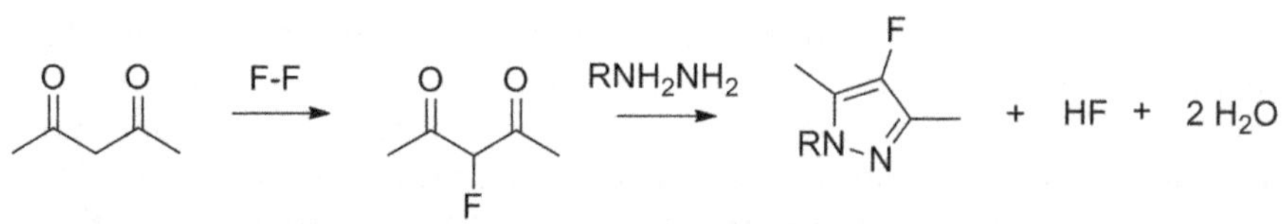

Scheme 12.2 An example of an atom-efficient electrophilic fluorination and hydrazination flow process with minimal by-product formation.[6]

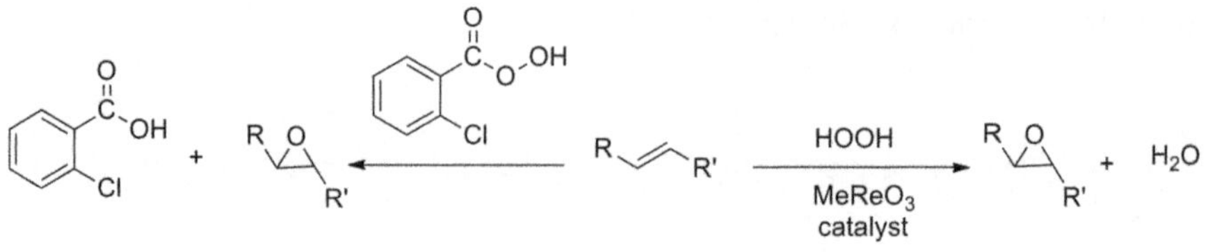

Scheme 12.3 Comparison of waste produced in an alkene epoxidation with MCPBA and hydrogen peroxide.

12.4 Principle 3: Wherever Practicable, Synthetic Methodologies Should be Designed to Use and Generate Substances That Possess Little or no Toxicity to Human Health and the Environment

The additional process control provided by continuous-flow technology means that non-toxic and environmentally benign reagents can be used that would be unsafe or impractical to handle in batch. Small volumes of toxic compounds can also be carefully generated *in situ* without the need for isolation. Seeberger recently developed a process for preparing large quantities of the anti-malarial drug artemisinin. This synthesis utilizes a synthetic transformation that is inaccessible using traditional batch methods due to safety and toxicity concerns.[8,9] There is a global shortage of artemisinin so many of the people who need it most in the developing world cannot access it.

A key step towards artemisinin from more readily available artemisinic acid is the formation of allylic hydroperoxide (Scheme 12.4). The peroxide product is generated using photochemistry and molecular oxygen. Oxygen and light are inexpensive, readily available and environmentally benign, and circumvent the use of toxic metal-based oxidants. Unfortunately, oxygen is highly flammable and achieving uniform light irradiation makes the process unsuitable for batch synthesis on a large scale. Continuous-flow techniques allow the addition of oxygen to be carefully controlled. Chlorinated solvents are usually necessary for the safe handling of oxygen, but recently the process has been modified to use toluene because only a small inventory of oxygen is present in the flow system at any one time. The synthesis route is fully telescoped, thereby avoiding accumulation of peroxide products, which are potentially explosive.

12.5 Principle 4: Design New Products That Preserve Functionality While Reducing Toxicity

Minimising toxicity, whilst maintaining functionality and efficacy, is a particularly difficult aspect in the design of safer chemicals. A way of reducing the impact of hazardous reagents is through their immediate synthesis and

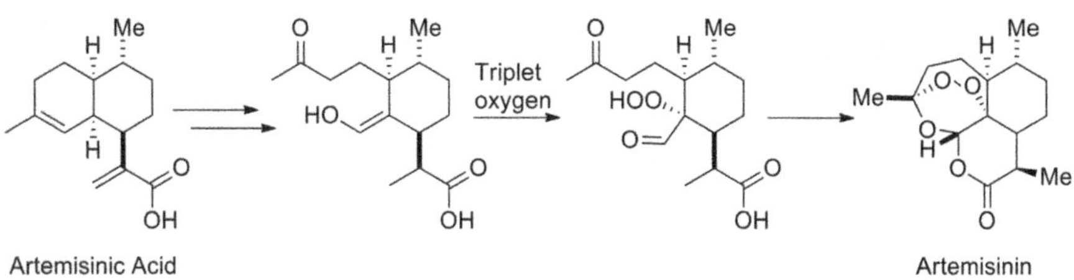

Scheme 12.4 Synthesis of artemisinin from artemisinic acid.

use in further reactions. This telescoping avoids the need to transport and stockpile large quantities, allowing for the production of the required amount on-demand.

The issue of potential genotoxic impurities (PGIs) in active pharmaceutical ingredients (APIs) and drug products continues to receive considerable attention. Alkyl halides are well-known to be potentially genotoxic and are, hence, a safety concern in medicinal chemistry.[10]

Flow chemistry can be used to form low concentrations of toxic compounds at any one time, which could potentially be reacted *in situ* to form more complex structures. Recent work by Kappe,[11] shows that benzylic bromides can be formed using *N*-bromosuccinimide (NBS) under photochemical conditions, in good yield, with minimal di-bromination occurring. The radical reactions were activated with a readily available household compact fluorescent lamp using a simple flow reactor design based on transparent fluorinated ethylene polymer (FEP) tubing. Additionally, the reactions were carried out using acetonitrile as the solvent, thus avoiding hazardous chlorinated solvents, such as CCl_4, which are commonly used in radical brominations (Scheme 12.5).

The formed benzyl bromides can then potentially be used in a range of telescoped chemical transformations, including alkylations. The flow synthesis of a number of receptor ligands, using benzyl bromides, has been demonstrated by Ulven.[12] Alternatively, C–C bond coupling reactions can be performed using flow electrosynthesis, as shown in Scheme 12.6.[13]

12.6 Principle 5: Minimise the Use of Auxiliary Reagents and Solvents

Flow processes offer potential for avoiding or minimizing solvent, because higher viscosities with more intense mixing and better energy management can be achieved than in batch.

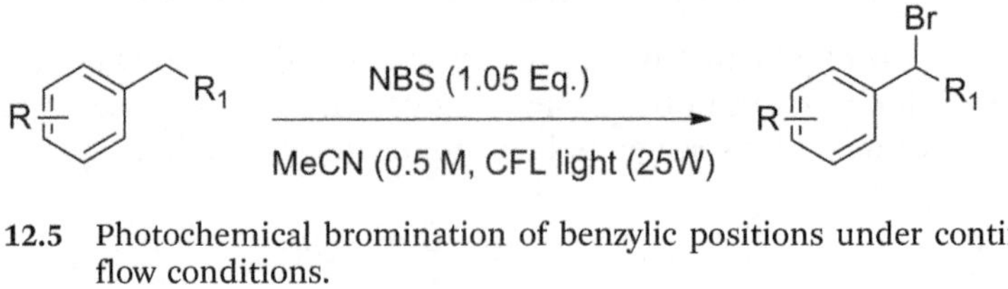

Scheme 12.5 Photochemical bromination of benzylic positions under continuous-flow conditions.

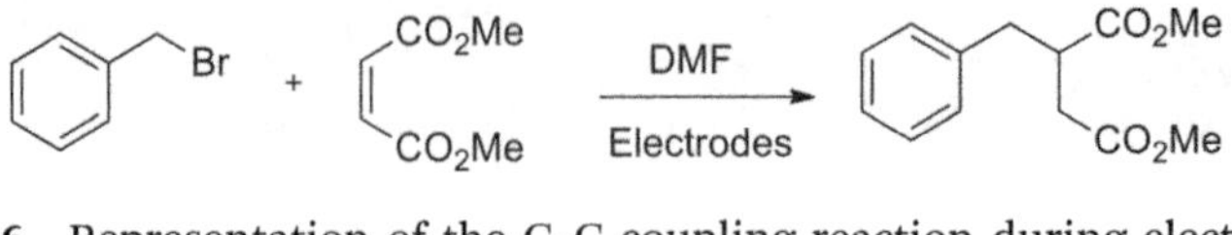

Scheme 12.6 Representation of the C–C coupling reaction during electrosynthesis in a microreactor.

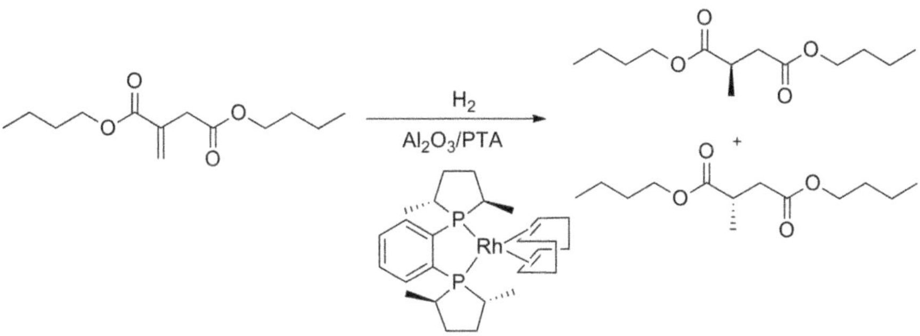

Scheme 12.7 Hydrogenation of dibutyl itaconate to give dibutyl 2-methylsuccinate.

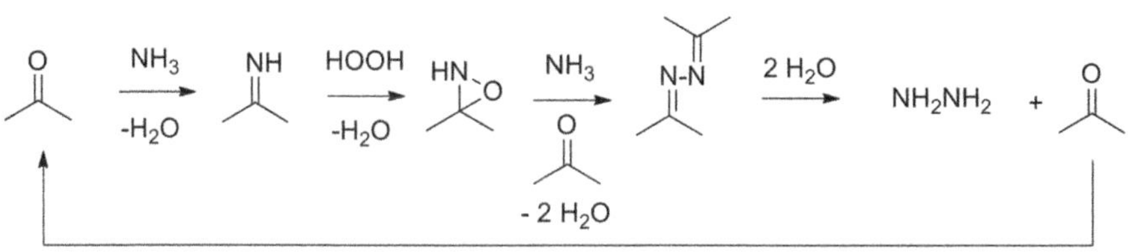

Scheme 12.8 Flow process to make hydrazine, illustrating minimal use of the auxiliary, acetone (separation and recycle), which is used to achieve a selective reaction.

An example of this is from Cole-Hamilton (Scheme 12.7).[14] He reported a flow process for the catalytic asymmetric hydrogenation of neat dibutyl itaconate in high yield and optical purity using Augustine's method for supporting rhodium MeDuPhos on alumina using a polyoxoanion. The product was produced directly without the need for purification at a rate of 3 g h^{-1}.[15]

Currently, there are only a few examples of minimizing reagents using flow methods. This can be achieved by using higher energies and shorter residence times to achieve more selective reactions. The Elf-Atochem continuous process to make hydrazine involves hydrogen peroxide oxidation of acetone imine (Scheme 12.8). After ring-opening the intermediate oxaziridine with ammonia–acetone to make *bis*-dimethyl hydrazone, hydrolysis yields the product hydrazine and the acetone auxiliary is regenerated and recycled.[16]

The use of solvents within product work-up usually represents the largest process waste. These include water or water-immiscible solvents for extraction, those for crystallization, solids washing, cleaning and flushing. Flow methods for product separation and isolation, such as continuous liquid–liquid extraction, crystallization, filtration, distillation and adsorption, offer the opportunity for more intense processes that minimize the use of work-up solvent.

12.7 Principle 6: Minimal Energy Requirements

Efficient use of energy and time is a key goal of green chemistry and reaction engineering.[17] The use of plug flow reactors enables reactions to be

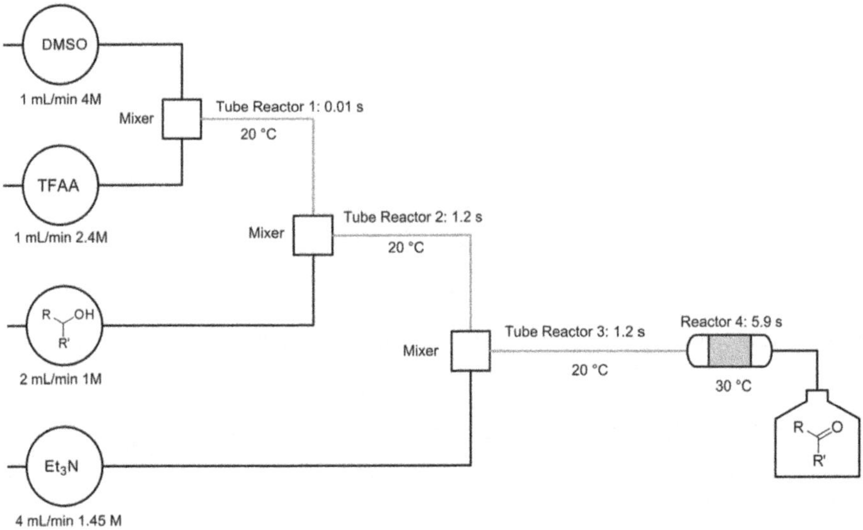

Figure 12.1 Simplified reactor schematic for the continuous Moffatt–Swern oxidation at room temperature with dimethyl sulfoxide (DMSO) and trifluoroacetic anhydride (TFAA) of a variety of alcohols (1-decanol 71%, 2-octanol 89%, cyclohexanol 88%, benzyl alcohol 75%).

conducted at temperatures and pressures far beyond those achievable in a batch system, which are typically limited to the atmospheric boiling point of the solvent. Reactions performed in flow are therefore often much more rapid than a corresponding batch process resulting in improved energy, time and space efficiency.

In addition, the increased surface area to volume ratio of plug flow reactors results in more efficient heat transfer. This enables reactions that are typically performed at cryogenic conditions in batch being safely operated at near ambient conditions in a flow reactor, resulting in a significant energy saving. As an example, Yoshida *et al.* demonstrated a Moffatt–Swern oxidation using a microreactor system at temperatures up to 20 °C, much higher than the conventional batch synthesis, which is operated below −50 °C.[18] This was accomplished by conducting the reactions in a series of tubular reactors and preventing the decomposition of unstable intermediates by using extremely short residence times (0.01 s), as shown in Figure 12.1.

12.8 Principle 7: Renewable Raw Materials

Biomass has been widely utilised throughout human history to provide important products including pharmaceutical products, dyes, flavours and fragrances. More recently, the conversion of biomass into fuels and chemicals has received considerable attention as an alternative to petrochemical feedstocks.

Flow processes offer potential advantages for the production of biobased materials and have been combined with numerous robust catalytic systems and alternative reaction media, including supercritical fluids and ionic liquids. In particular, flow processing enables access to short residence times and high-energy regimes, which are commonly required for the reduction of highly oxidised bio-based feedstocks and the breakdown of lignocellulose biopolymers.

For example, the reduction of furfural, a biorenewable platform chemical, which is produced industrially (*ca.* 250 000 tonnes per annum) was performed by Stevens and co-workers using two flow reactors in series using supercritical CO_2. They used a switchable temperature-sensitive catalyst system; the first reactor contained copper chromite, which at moderate temperatures (*ca.* 120 °C) would selectively reduce the carbonyl group to an alcohol or at elevated temperatures (*ca.* 240 °C) was able to fully reduce the carbonyl group to a methyl group. The second reactor contained Pd/C, which at elevated temperatures hydrogenated the furan ring or decarbonylated furfural to furan in the absence of hydrogen. This enabled five different products to be synthesized from furfural in high yields (Figure 12.2) with rapid switching between the products possible.[19] Note however that the use of copper chromite, as a class 1 carcinogen, is undesirable, breaking Principle 3.

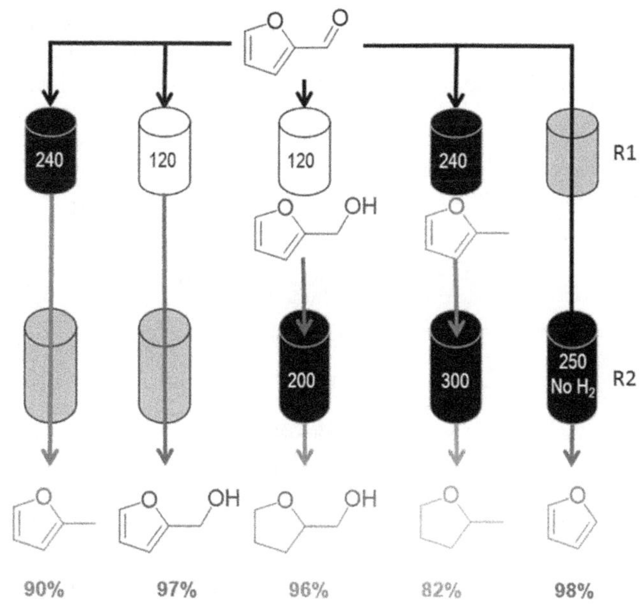

Figure 12.2 Simplified schematic showing switchable production of a variety of chemical products from furfural using a twin series reaction system. R1 = copper chromite catalyst, R2 = palladium on carbon.

12.9 Principle 8: Avoid Unnecessary Derivatization

The use of protecting groups has been essential in the isolation of many important organic molecules. Protection, however, increases the number of reaction steps and reduces atom economy, which, in turn, reduces the greenness of the process. Flow reactors are now being used in a range of protecting-group-free syntheses.

The flow synthesis of a Macbecin I precursor has been demonstrated by Yoshida.[20] Under batch conditions, the nitro-aromatic is reduced to the amine, then protected, the aryllithium then generated and reacted with an aldehyde. Finally, the amine is deprotected to give the product (Scheme 12.9).[21]

Alternatively, the microreactor flow synthesis allows the use of unprotected, functionalised organolithium reagents by using residence time control of flow microreactor systems (Scheme 12.10). A reduction of the nitro group should give the Macbecin I precursor, avoiding any protection–deprotection processes.[†]

Similarly, Booker-Milburn demonstrated that (±)-neostenine can be synthesised without the use of protecting groups, with the fused pyrrolo[1,2-*a*]azepine core being isolated by a [5 + 2] maleimide photocycloaddition using a custom FEP flow reactor (Scheme 12.11).[22]

12.10 Principle 9: Catalytic Reagents are Superior to Stoichiometric Reagents

Catalytic processes can prevent waste and circumvent the need for recycling processes. Catalytic systems often reduce the energy requirements for a process because milder operating conditions can be used and no separation

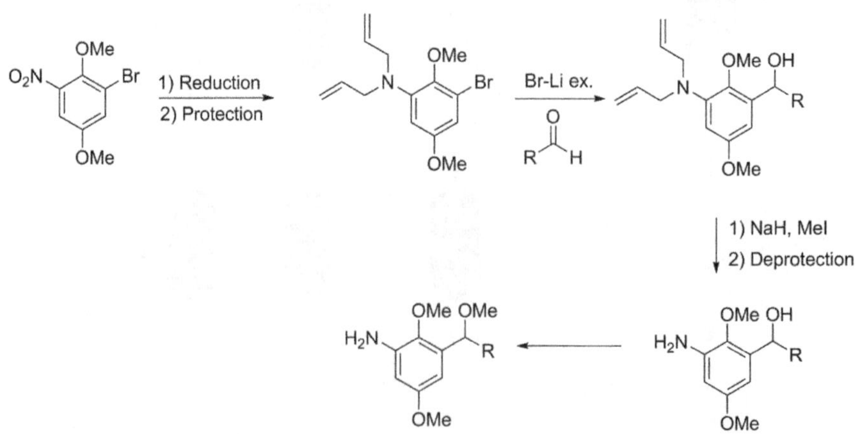

Scheme 12.9 Synthesis of a Macbecin I precursor under batch conditions.

[†]The solvent HMPA used in this process is a suspected human carcinogen, and therefore an undesirable solvent whose use here breaks Principle 3.

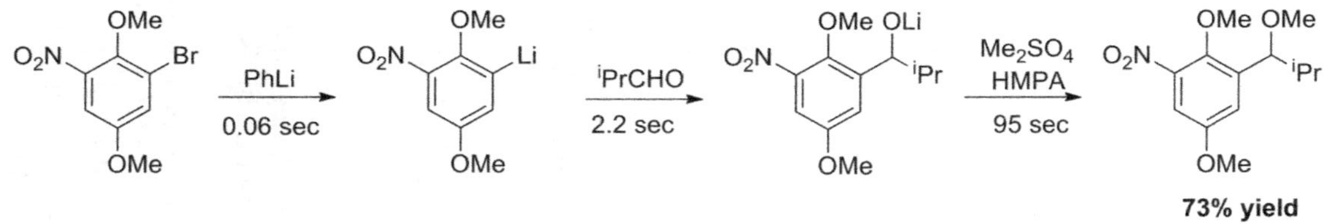

Scheme 12.10 Synthesis of a Macbecin I precursor under flow conditions.

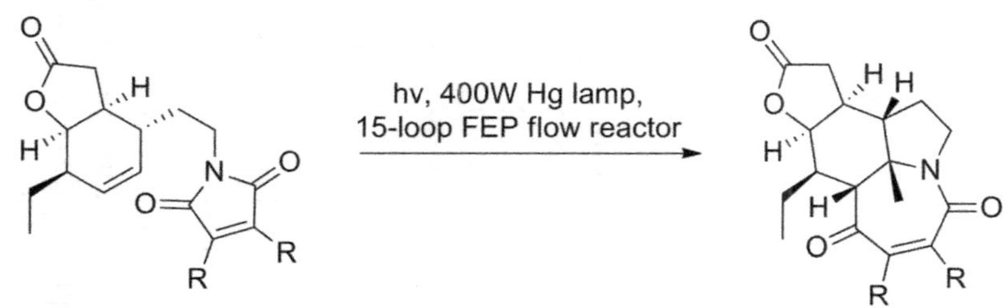

Scheme 12.11 Protecting group-free synthesis of (±)-Neostenine.

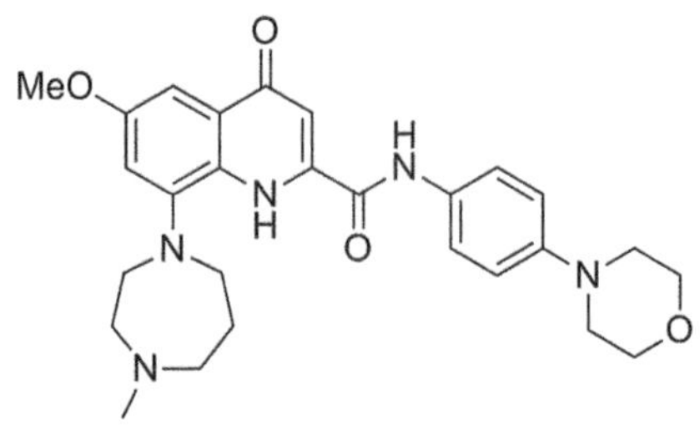

Figure 12.3 5-HT_{1B} antagonist quinolone derivative.

processes are necessary post reaction when a highly selective catalyst is used. Immobilized catalysts are often utilized in flow processes to simplify the purification and to enable multi-step reactions without the need for isolation.[23] Ley and co-workers reported a continuous-flow process that utilized a solid-supported Pd/C catalyst and subsequently a scavenger column to remove any leached palladium. The 5-HT_{1B} antagonist quinolone derivative was prepared with 18% overall yield and higher than 98% purity.[24] This was a significant improvement to a previously published batch synthesis by AstraZeneca, which gave the quinolone compound in 7% overall yield, (Figure 12.3).[25]

As mentioned previously, in Principle 5, Cole-Hamilton recently reported a very efficient, solvent-free hydrogenation of the prochiral substrate dibutyl itaconate to give the single enantiomer (*S*)-dibutyl 2-methylsuccinate in a plug flow reactor (Scheme 12.7) using ruthenium catalysis.[14] The catalyst was supported onto alumina and phosphotungstic acid (PTA), a heteropoly acid, was used as an anchoring agent for the Rh metal, which ensured Rh leaching was minimal. Supporting the metal catalyst prevented a drop in conversion over time and stopped catalyst recovery from being required.

12.11 Principle 10: Design New Products With Biodegradable Capabilities

Chemical products should be designed so that at the end of their function they break down into mild degradation products *via* biodegradation, hydrolysis or photolysis and do not persist in the environment. This is also true for the solvents used in a process.

The synthesis of aryl sulfonamides has been demonstrated under flow conditions using biodegradable PEG as a solvent (Scheme 12.12).[26] PEG 400 was chosen because it is relatively cheap, readily recyclable, biodegradable and water-soluble, making it easily removable by simple workup operations. It also has suitable viscosity for the required flow conditions.

12.12 Principle 11: Real-time and Online Process Analysis

Flow reactors typically enable simple integration of various parameter probes, such as temperature, pressure, conductivity and pH. In addition, due

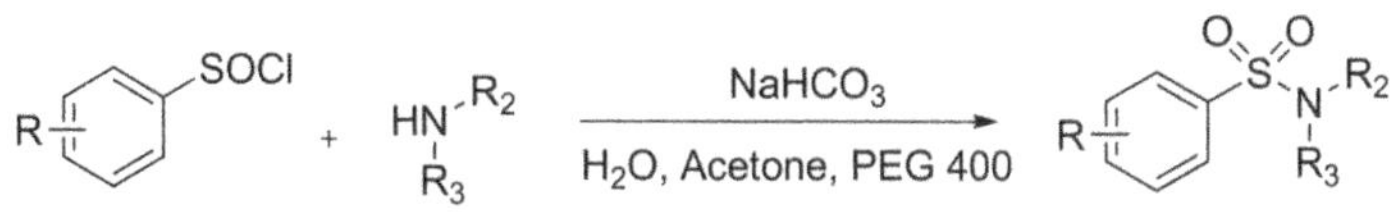

Scheme 12.12 The synthesis of sulfonamides using biodegradable PEG as the solvent.

to the considerable interest in the automation of data collection, several online methods of sampling have been developed, enabling rapid data collection of both spectroscopic and chromatographic techniques.

More recently, self-optimising processes have been developed, in which reactor control, on-line analytical techniques and optimisation algorithms have been combined into a single automated system to deliver optimised product streams without requiring intervention from the operator. This strategy enables reactions to be fully optimised rapidly and provides a facile route to consider all operating parameters rather than a univariate approach, which is commonly used.

Jensen *et al.* demonstrated an integrated microreactor system with online HPLC analysis to optimise the Heck reaction of 4-chlorobenzotrifluoride and 2,3-dihydrofuran (Figure 12.4).[27] The optimisation was controlled using a Nelder–Mead Simplex method, a 'black-box' approach, which required no *a priori* reaction or gradient information. They demonstrated that this process could be optimised to produce a yield of 83% after 19 experiments, each taking approximately 20 minutes including analysis time.

Jumbam and co-workers also showed that it was possible to optimise using green reaction metrics for the methylation of *n*-pentanol with dimethyl carbonate using a γ-alumina catalyst (Scheme 12.13).[28] In this example, operating conditions were found by optimizing for yield, space–time yield, E factor (with and without solvent) and a weighted yield function. This resulted in either 97.6% yield, a space–time yield of 42.9 kg L^{-1} h^{-1} and *E* factors of 1.4 and 3.3 (including solvent) and a more balanced approach with a yield of 80% and a space–time yield of 37.2 kg L^{-1} h^{-1} when optimising for the weighted yield.

12.13 Principle 12: Substances should be chosen so as to Minimize the Potential for Chemical Accidents, Including Releases, Explosions and Fires

Continuous processing is beneficial for handling hazardous intermediates because only a small inventory is produced at any one time. The control of exothermic reactions is also improved due to the enhanced heat transfer properties provided by the high surface area to volume ratio. Diazomethane (DZM) is a versatile C1 building block that gives direct access to valuable organic molecules (Scheme 12.14). Reactions involving DZM are generally atom-efficient and often give nitrogen gas as the only by-product which is

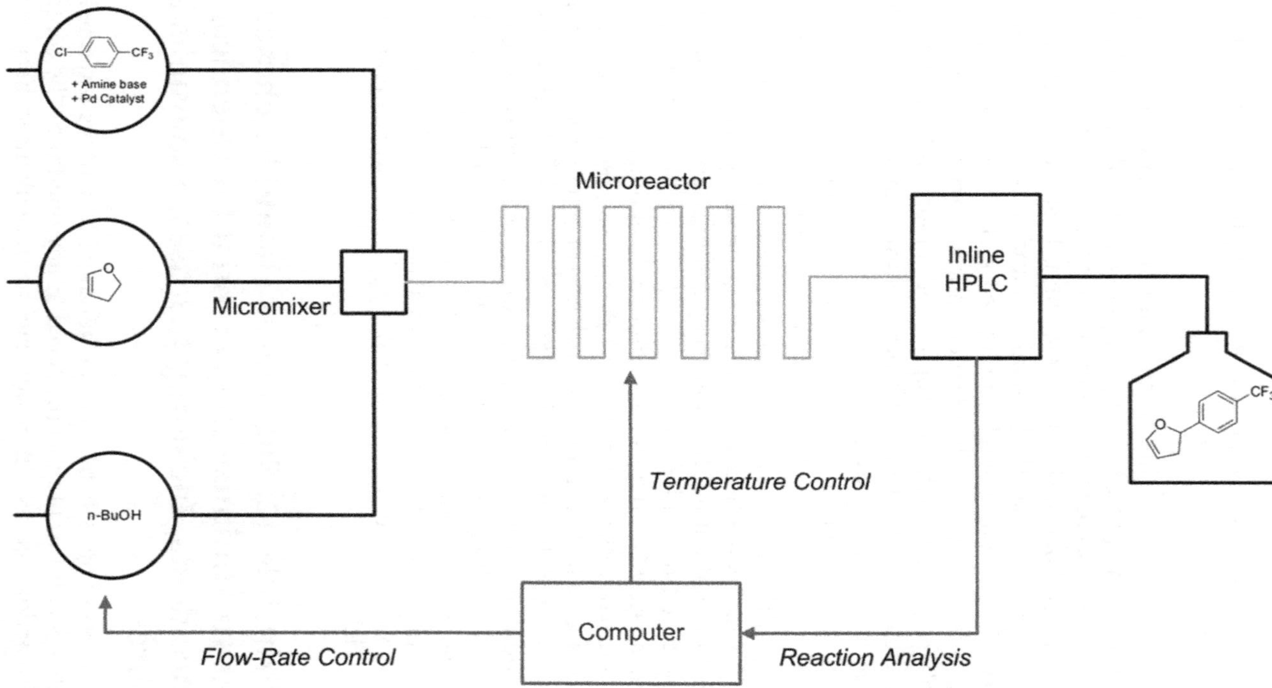

Figure 12.4 Schematic of self-optimising microreactor system for the Heck reaction with inline HPLC analysis. Reprinted with permission from Wiley.[27]

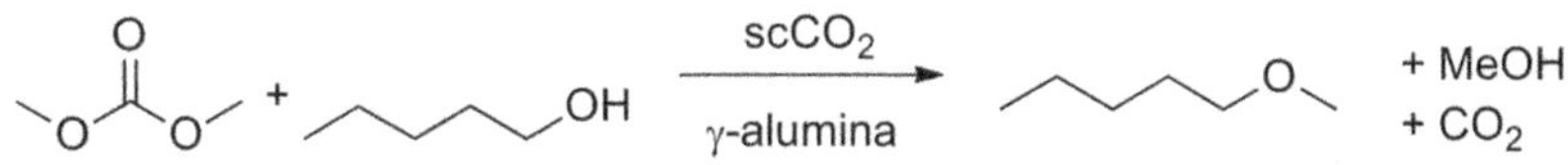

Scheme 12.13 Methylation of n-pentanol using dimethyl carbonate as a sustainable methylation reagent and a γ-alumina catalyst.

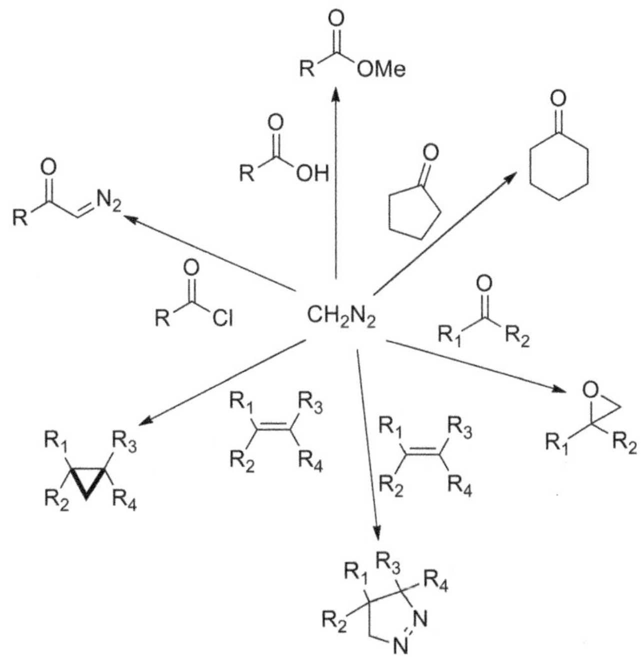

Scheme 12.14 Reactions utilizing diazomethane as a reagent.

environmentally benign and easy to remove.[29] Unfortunately DZM is underutilized in organic synthesis due to safety concerns. DZM is an odourless, highly toxic (PEL = 0.2 ppm) and explosive yellow gas (bpt. = −23 °C).

Kim recently reported a dual-channel microreactor system for the use of diazomethane in continuous flow.[30] The two channels are separated by a thin hydrophobic poly(dimethylsiloxane) (PDMS) membrane. DZM is generated in the aqueous layer by reacting KOH with *N*-methyl-*N*-nitroso-*p*-toluenesulfonamide (Diazald) and then selectivity diffuses through the membrane to the organic layer where it reacts with the substrate. DZM is generated and consumed at a steady rate, which provides an improved safety profile (Scheme 12.15). *In situ* generation of DZM avoids its storage and manual handling. The membrane simplifies downstream purification steps. DZM is contained within an end-to-end fully integrated continuous flow process. Phoenix Chemicals have developed an industrial scale process that can produce 80 tonnes of DZM per annum and only 80 g inventory of DZM is generated at any one time.[29] Continuous processing gives access to

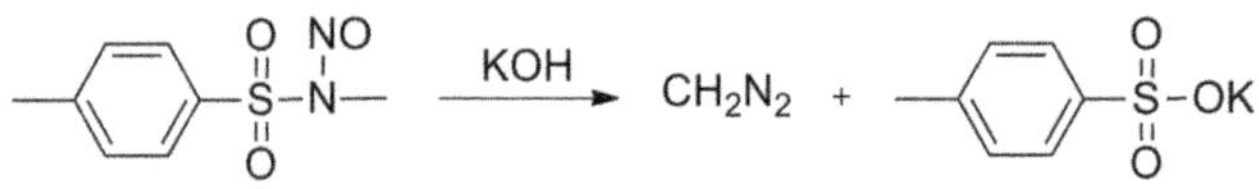

Scheme 12.15 Generation of diazomethane from Diazald.

chemistry that is green but would previously be avoided because of safety concerns.

12.14 Summary

- This chapter considers the beneficial combination, and potential impact, of continuous flow methodology with the twelve Principles of Green Chemistry as developed by Anastas.
- Whilst the application of continuous flow reactions is changing the way medicinal chemists are making and trying to identify lead bioactive compounds, the main impact of flow methods on green chemistry is in process development and manufacture, where the volumes of materials produced are far greater.
- The least wasteful strategy for synthesizing a product is one in which no by-product is formed and the reaction is 100% atom-efficient.
- A way of reducing the impact of hazardous reagents is through their immediate synthesis and use in further reactions.
- Flow processes offer potential for avoiding or minimizing solvent, because higher viscosities with more intense mixing and better energy management can be achieved than in batch.
- Chemical products should be designed so that at the end of their function they break down into mild degradation products *via* biodegradation, hydrolysis or photolysis and do not persist in the environment.
- Flow processes offer potential advantages for the production of bio-based materials and have been combined with numerous robust catalytic systems and alternative reaction media, including supercritical fluids and ionic liquids.

References

1. P. T. Anastas and M. M. Kirchhoff, *Acc. Chem. Res.*, 2002, **35**, 686.
2. B. M. Trost, *Angew. Chem. Int. Ed.*, 1995, **34**, 259.
3. C. A. Tolman, R. J. McKinney, W. C. Seidel, J. D. Druliner and W. R. Stevens, *Adv. Catal.*, 1985, **33**, 1.
4. G. W. Parshall and S. D. Ittel, *Homogeneous Catalysis: The Applications and Chemistry of Catalysis by Soluble Transition Metal Complexes*, Wiley-Interscience, New York, Second edn, 1992.
5. I. I. Junior, M. C. Flores, F. K. Sutili, S. G. F. Leite, L. S. M. de Miranda, I. C. R. Leal and R. O. M. A. de Souza, *Org. Process Res. Dev.*, 2012, **16**, 1098.

6. J. R. Breen, G. Sandford, D. S. Yufit, J. A. K. Howard, J. Fray and B. Patel, *Beilstein J. Org. Chem*, 2011, **7**, 1048.
7. M. C. A. van Vliet, I. W. C. E. Arends and R. A. Sheldon, *Chem. Commun.*, 1999, 821.
8. F. Lévesque and P. H. Seeberger, *Angew. Chem., Int. Ed.*, 2012, **51**, 1706.
9. D. Kopetzki, F. Lévesque and P. H. Seeberger, *Chem. - Eur. J.*, 2013, **19**, 5450.
10. D. P. Elder, A. M. Lipczynski and A. Teasdale, *J. Pharm. Biomed*, 2008, **48**, 497.
11. D. Cantillo, O. de Frutos, J. A. Rincon, C. Mateos and C. O. Kappe, *J. Org. Chem.*, 2014, **79**, 223.
12. T. P. Petersen, A. Ritze and T. Ulven, *Org. Lett.*, 2011, **11**, 5134.
13. P. He, P. Watts, F. Marken and S. J. Haswell, *Angew. Chem., Int. Ed.*, 2006, **45**, 4146.
14. R. Duque, P. J. Pogorzelec and D. J. Cole-Hamilton, *Angew. Chem., Int. Ed.*, 2013, **125**, 9987.
15. R. Augustine, S. Tanielyan, S. Anderson and H. Yang, *Chem. Commun.*, 1999, 1257.
16. L. D. Pesce and W. R. Jenks, *Riegel's Handbook of Industrial Chemistry*, ed. J. A. Kent, Synthetic Nitrogen Products, Kluwer, New York, Tenth edn, 2003, pp. 1157–1234.
17. J. B. Zimmerman and P. Anastas, *Environ. Sci. Technol.*, 2003, **37**, 94A–101A.
18. T. Kawaguchi, H. Miyata, K. Ataka, K. Mae and J. Yoshida, *Angew. Chem., Int. Ed.*, 2005, **44**, 2413.
19. J. G. Stevens, R. A. Bourne, M. V. Twigg and M. Poliakoff, *Angew. Chem., Int. Ed.*, 2010, **49**, 8856.
20. J. Yoshida, H. Kim and A. Nagaki, *ChemSusChem*, 2011, **4**, 331.
21. J. K. Belardi and G. C. Micalizio, *Angew. Chem., Int. Ed.*, 2008, **47**, 4005.
22. M. D. Lainchbury, M. I. Medley, P. M. Taylor, P. Hirst, W. Dohle and K. Booker-Milburn, *J. Org. Chem.*, 2008, **73**, 6497.
23. G. Jas and A. Kirschning, *Chem. - Eur. J.*, 2003, **9**, 5708.
24. Z. Qian, I. R. Baxendale and S. V. Ley, *Synlett*, 2010, 505.
25. C. L. Horchler, J. P. McCauley Jr., J. E. Hall, D. H. Snyder, W. Craig Moore, T. J. Hudzik and M. J. Chapdelaine, *Bioorg. Med. Chem.*, 2007, **15**, 939.
26. A. Gioiello, E. Rosatelli, M. Teofrasti, P. Filipponi and R. Pellicciari, *ACS Comb. Sci*, 2013, **15**, 235.
27. J. P. McMullen, M. T. Stone, S. L. Buchwald and K. F. Jensen, *Angew. Chem., Int. Ed.*, 2010, **122**, 7230.
28. D. N. Jumbam, R. A. Skilton, A. J. Parrott, R. A. Bourne and M. Poliakoff, *J. Flow Chem.*, 2012, **2**, 24.
29. L. D. Proctor and A. J. Warr, *Org. Process Res. Dev.*, 2002, **6**, 884.
30. R. A. Maurya, C. P. Park, J. H. Lee and D-P. Kim, *Angew. Chem., Int. Ed.*, 2011, **50**, 5952.

CHAPTER 13

Green Catalytic Direct Amide Bond Formation

FARHANA K. FERDOUSI AND ANDREW WHITING*

Centre for Sustainable Chemical Processes, Department of Chemistry, Durham University, Science Laboratories, South Road, Durham DH1 3LE, UK
*Email: andy.whiting@durham.ac.uk

13.1 Introduction

The amide bond (–CO–NH–) is arguably one of the most important groups in chemistry, particularly in peptide synthesis.[1] The ubiquity and importance of amide bonds in biology has led to them being present in many active pharmaceutical ingredients (API) and in a huge number of *de novo* drug molecules,[2] like antibiotics, antivirals, immuno-modulators, anti-angiogenesis agents, central nervous system (CNS) and neuro-active agents, anti-obesity drugs, and many other therapeutic areas (Figure 13.1).[3]

Over the last ten years, the number as well as the capacity of bulk peptide producers and peptide synthesis companies has been growing significantly throughout the world. For cancer treatment, there are 218 peptide drugs being developed in different projects.[4] *Comprehensive Medicinal Chemistry* reports that around 25% of all synthetic pharmaceuticals contain at least one amide bond. In addition to pharma drugs, the amide bonds also appear as one of the most prolific moieties in agrochemicals, natural products and fine chemicals.

RSC Green Chemistry No. 46
Green and Sustainable Medicinal Chemistry: Methods, Tools and Strategies for the 21st Century Pharmaceutical Industry
Edited by Louise Summerton, Helen F. Sneddon, Leonie C. Jones and James H. Clark

Published by the Royal Society of Chemistry, www.rsc.org

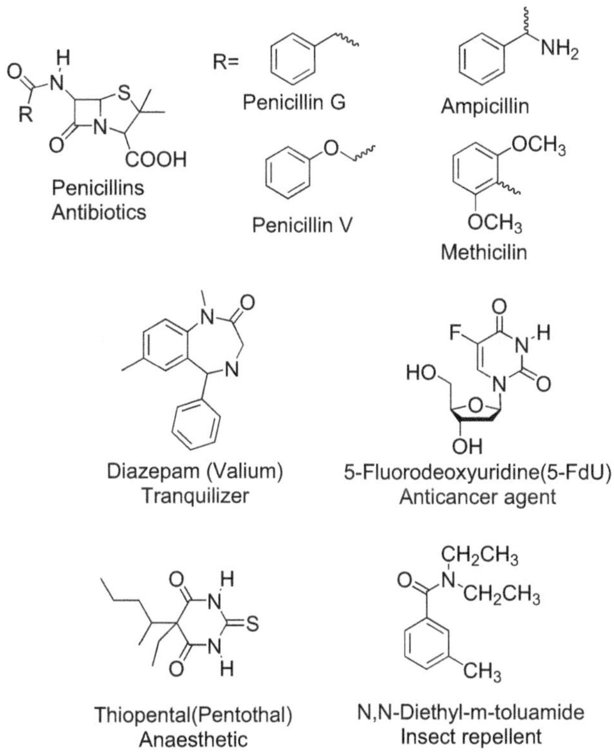

Figure 13.1 Some important amide-containing drugs.

13.2 Current Industrial Methods for Synthesizing Amide Bonds

According to data from medicinal chemistry campaigns, 16% of all reactions performed in the synthesis of pharmaceuticals (containing amide bonds) involve acylation of amines with activated carboxylic acids.[5] In a recent industry-led survey of 128 drug syntheses, acylations accounted for 12% of the reactions (second only to heteroatom alkylations/arylations), of which 66% were *N*-acylations to form amides.[6]

This activation of carboxylic acids is generally carried out using different coupling agents (Figure 13.2). Today, the majority of peptide drugs of 13–15 amino acids are synthesized using different coupling reagents, especially involving N-Boc chemistry and conventional DCC/HOBt methods.[2] Recently, a shift of industry to solid-phase Fmoc chemistry has increased the use of modern coupling techniques (HBTU/TBTU and HCTU/TCTU).

Despite accounting for a large proportion of the transformations, none of the *N*-acylation reactions used for amide bond formation were catalytic. Current methods used for amide formations are remarkably general, but

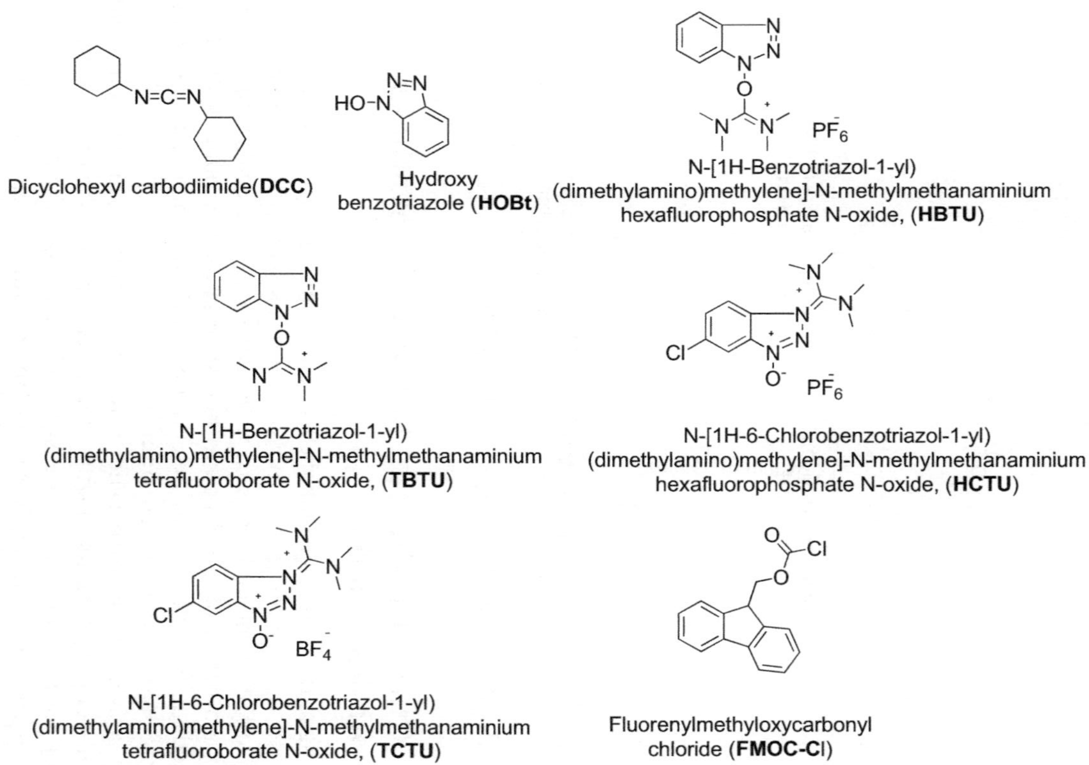

Figure 13.2 Some modern coupling agents used in industry for amide bond formation.

at the same time, widely regarded as expensive and unrefined. This is because stoichiometric use of all of these coupling agents and activating agents is atom uneconomic, and there are issues with product purification and toxic reactive reagents. Furthermore, even the best stoichiometric reagents often fail for the synthesis of sterically hindered amides,[6] and hence have caused a rethink in direct amide formation strategies. Therefore, these types of reagents might ideally be exchanged for other processes,[7] including microwave[8] and new thermal catalytic processes,[9] introducing cheaper reagents with reduced impact on the environment.[10]

Not surprisingly, in 2007 the American Chemical Society Green Chemistry Institute (comprising members from major pharmaceutical industries worldwide) voted 'amide formation avoiding poor atom economy reagents' as the top challenge for organic chemistry.[10] In a separate survey, concerned with identifying key green chemistry research areas, amide formation avoiding poor atom economy reagents was voted the highest priority reaction that companies would like to see investigated.[10] Both these observations highlight the need for new, atom-efficient, economic and preferably environmentally friendly catalytic processes for synthesising amide bonds to be developed.

13.3 Research Trends: The Development of New Catalytic Systems

In recent years, several effective catalytic systems have been developed, which has led to the direct transformation of carboxylic acids (**1**) and amines (**2**) to form amides (**3**), with water as the only side-product (Scheme 13.1).

The most studied and perhaps promising catalytic systems in direct amidation so far are boric and boronic acid-based systems. However, other catalytic systems, such as triazines and other metal-based catalysts, have also been found to be impressive.[1,11] In addition, heterogeneous catalysis is also possible in this area, with the application of microwave conditions.[6]

Since 2005, boric acid has been used[12] as an inexpensive, non-toxic, environmentally neutral (risk phrases associated with boric acid: R62, R63; LC_{50} is >1000 to fish and to aquatic invertebrates is 133–226 mg kg^{-1})[13,14] and effective catalyst for direct amidation. Excellent catalytic activity was found for the condensation of simple aliphatic and benzyl amines and carboxylic acids, such as 4-phenylbutyric acid and benzylamine with a

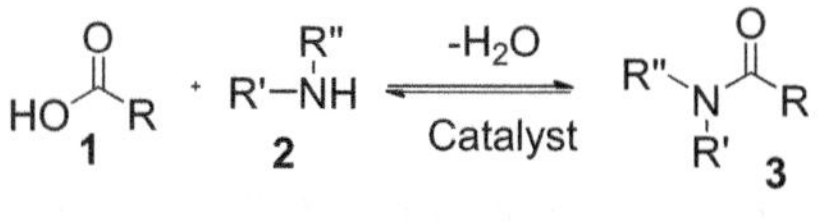

Scheme 13.1 Catalytic direct amide formation.

catalyst loading of 5 mol% (eqn (13.1)), giving 99% yield of amide (refluxing in heptanes; b.p. ~98 °C) with water removal accomplished azeotropically.

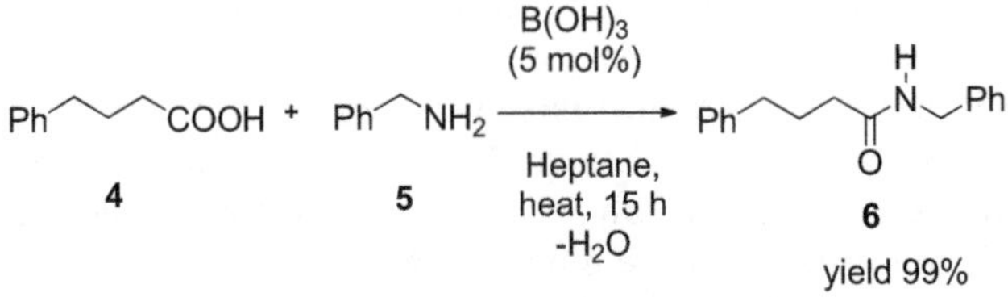

(13.1)

Similarly, an electron withdrawing substituent attached to aniline was also found to be effective, but higher catalyst loadings (up to 25 mol%, *i.e.* due to the less nucleophilic 3,5-dimethylpiperidine amine being used) gave 95% of amide while refluxing in toluene (b.p. ~110 °C).[7]

Interestingly, it is evident that the reactivity of boric acid is better at high temperatures, and it declines at lower reaction temperatures.[12,15] Subsequently, a wide variety of boron-based reagents, such as catechol and arylboronic acid-based catalysts, have been developed for direct amide formation. The enhancement of the Lewis acidity of boron has been proven to enhance catalytic activity (*e.g.* chlorocatechol derivative **9** or **10**), as in eqn (13.2).[16,17]

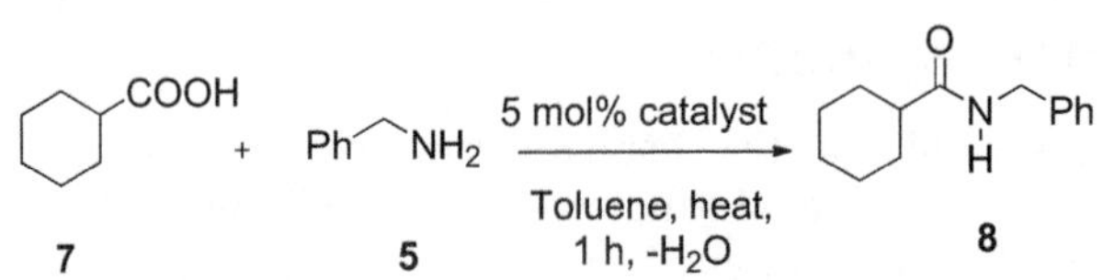

Catalyst	Yield (%)
$B(OH)_3$	2
9	52
10	62
11	32

(13.2)

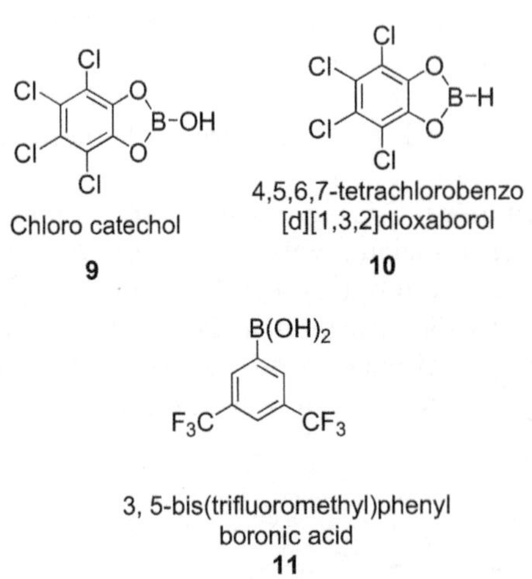

The use of chlorocatechol borate (**9**) catalyst is advantageous compared with 3,5-bis(trifluoromethyl)phenylboronic acid (**10**), being less sterically hindered, which increases the Lewis acidity. However, for the more sterically hindered acids and amines (such as cyclohexanecarboxylic acid **7**) catalyst **9** is more efficient than **10** (conversion is 52% and 32%, respectively) (eqn (13.2)). The use of 5 mol% of 4,5,6,7-tetrachlorobenzo[*d*][1,3,2]dioxaborol (**11**) in this reaction afforded higher yield (62%) than for the other catalysts.[17]

The area of boronic acids (Figure 13.3) as catalysts has attracted considerable interest since 1996, after Ishihara *et al.* reported their first use in the amidation of carboxylic acids using amines with catalyst **10**.[18] These reactions required refluxing toluene, xylene or mesitylene conditions, with molecular sieves as a drying agent. Successive reports incorporated the benefits of recyclable boronic acid **12** (Figure 13.3) as an effective amidation catalyst that can be obtained in a fluorous phase at the end of the reaction.[19] Pyridiniumboronic acid **13** is also an effective boronic acid catalyst that can be applied in relatively polar organic solvents.[20]

Recently, effective catalytic amidation has been examined at reduced reaction temperatures. Tertiary amine-containing boronic acid **16** catalyses several amidation reactions in fluorobenzene under reflux (85 °C).[21] In a recent study, 3,4,5-trifluorophenylboronic acid (**13**) and *o*-nitrophenylboronic acid (**19**) were found to be the best catalysts at low temperature (65–68 °C) for amidations as well as peptide and dipeptide formation (though higher catalyst loadings were required for slower dipeptide formations).[22] This method conserved the enantiopurity of the chiral starting material in most cases. An interesting synergistic catalytic effect

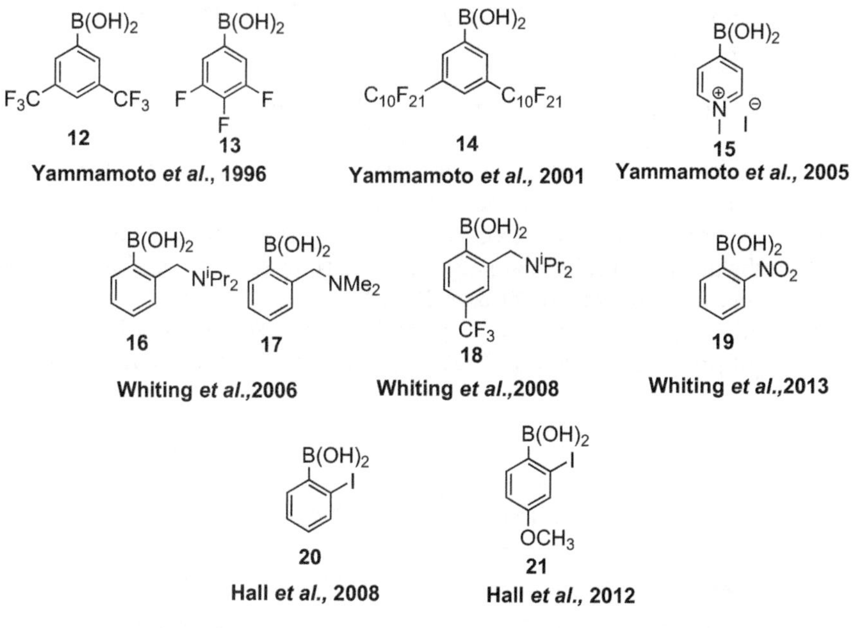

Figure 13.3 Some important aryl boronic acid catalysts used for the synthesis of amide bonds.

between two arylboronic acids (**11** and **17**) was also discovered.[23] *Ortho*-iodobenzene boronic acid (**20**) is able to catalyse amide formation at room temperature, in the presence of molecular sieves as a dehydrating agent.[23] The electron rich variant **21** was also recently revealed to show even greater sensitivity towards amide formation.[24] Above all, the reactivity of substrates in these boron-mediated amidation processes follows a similar trend to their reactivity in thermal amidation reactions.[7]

13.4 Pros and Cons of New Catalytic Methods

Stoichiometric activation of carboxylic acid has been generally employed, resulting in the production of substantial amount of by-products.[1,2,6] On the contrary, catalytic direct amide formation methodology has major advantages of atom economy and avoids protection/deprotection steps. Boric acid catalyzed direct amide formation exploits all these advantages in a potentially green and sustainable setting. Although this approach is very attractive due to the low cost of this compound, moderate to high reaction temperatures are required and the substrate scope is limited in comparison with boronic acid catalyzed processes. However, long reaction times, narrow solvent compatibility, lack of general substrate scope and limited catalytic activities are also associated with other aryl boronic acid derivatives.

13.5 Application and Outlook for Catalytic Amidation in Pharmaceuticals and Fine Chemicals

In recent years, boric acid has been used in direct amidation from laboratory scale to the industrial proportion. Thus, the syntheses of some important drugs have been explored by a number of pharmaceutical laboratories recently. From 2006, GSK synthesized efaproxiral (**23**)[25] using 7.7 mol% of boric acid loading in the direct amidation of 4-hydroxyphenylacetic acid and 3,5-dimethylaniline in toluene using Dean–Stark water removal for 18 hours (86–95% yield), as shown in Scheme 13.2.[26]

Dr Reddy's Laboratories has reported more detailed studies on the synthesis of pharmaceutical intermediates using direct amidation technology

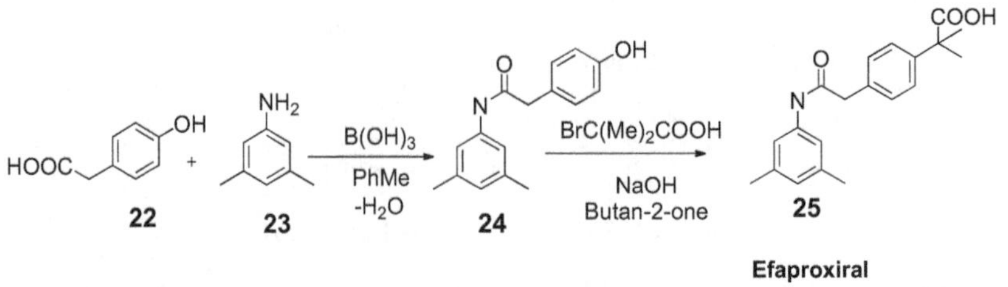

Scheme 13.2 Synthesis of efaproxiral.

in the literature. Seven APIs were investigated using boric acid and phenyl boronic acid as catalysts. Galanthamine was prepared in an impressive yield (86–93%) using 10 mol% of boric acid in the direct amidation process in refluxing toluene or xylene. Another striking application was alfuzosin intermediate synthesis by the same procedure (yield: 94–95%) but over a longer reaction time (6 hours). Since 2003, alfuzosin has been used as a treatment for benign prostatic hyperplasia. Dr Reddy's Laboratories has also been successful in synthesizing the antidiabetic agent, repaglinide (yield 72%) and an antiarrhythmic agent, flecainide (yield: 87%) with a high purity using the same catalyst loading in the direct amidation.

Pfizer has also been investigating catalytic direct amidation reactions. One of the most encouraging substrates is the synthesis of zamifenacin, which has been used for overactive bladder and irritable bowel syndrome. Only 2 mol% boric acid catalyst loading gave complete conversion of amide with Dean–Stark water removal in the direct amidation. Unfortunately, in 1990, zamifenacin was put on hold and replaced by darifenacin, so the synthesis of zemifenacin was never scaled up beyond laboratory scale.

The industrial application of catalytic direct amide formation is still little explored and the current processes are still unoptimised. Moreover, due to the high tolerance to water, acids or bases, boron-based catalysts could be considered one of the most potentially interesting areas of research in the arena of direct amidation used in pharmaceuticals and fine chemicals. Clear steps have been followed so far by the scientific community in this direction for amide transformation, especially with boron-based catalysts,[16,17,21,23] and a renaissance in this area has already been triggered.

13.6 Conclusion

Catalytic direct amide formation is important for pharmaceuticals or fine chemicals as it leads to more cost effective and greener processes with extraordinarily high atom economy. Most of the impressive catalytic processes have been developed with boric acid and aryl boronic acid based catalysts. Current industrial methods for amide bond formation mostly involve different coupling agents, like carbodiimides. Conventional coupling methods for amide bond formation suffer from complicated product isolations and poor atom economy in comparison with modern catalytic direct amidation technology. Several challenges remain for green catalytic direct amidation.

References

1. C. A. G. N. Montalbetti and V. Falque, *Tetrahedron*, 2005, **61**, 10827–10852.
2. O. Marder, F. Alberico, *Chemica OGGI - Chemistry Today*, 2003.
3. A. Loffet, *J. Peptide Sci.*, 2002, **8**, 1–7.

4. http://www.researchandmarkets.com/reports/1196702/peptides_in_oncology_drug_pipeline_update_2013.
5. S. D. Roughley and A. M. Jordan, *J. Med. Chem.*, 2011, **54**, 3451–3479.
6. V. R. Pattabiraman and J. W. Bode, *Nature*, 2011, **480**, 471–479.
7. H. Charville, D. Jackson, G. Hodges and A. Whiting, *Chem. Commun.*, 2010, **46**, 1813–1823.
8. L. Perreux, A. Loupy and F. Volatron, *Tetrahedron*, 2002, **58**, 2155–2162.
9. J. W. Comerford, T. J. Farmer, D. J. Macquarrie, S. W. Breeden and J. H. Clark, *ARKIVOC*, 2012, **vii**, 282–293.
10. D. J. C. Constable, P. J. Dunn, J. D. Hayler, G. R. Humphrey, J. L. Leazer, R. J. Linderman, K. Lorenz, J. Manley, B. A. Pearlman, A. Wells, A. Zaks and T. Y. Zhang, *Green Chem.*, 2007, **9**, 411–420.
11. B. M. Monks and A. Whiting, Direct amide formation avoiding poor atom economy reagents, *Sustainable Catalysis* published by Wiley and Sons, Inc., Hoboken, New Jersey, 2013, pp. 89–110.
12. P. Tang, *Org. Synth.*, 2005, **81**, 262–272.
13. *The Merck Index - an Encyclopedia of Chemicals Drugs and Biological*, ed. S. Budavari, Merck and Co., Whitehouse Station, NJ, 1996.
14. National Pesticide Information Centre, 1998.
15. K. Arnold, B. Davies, R. Giles, C. Grosjean, G. Smith and A. Whiting, *Adv. Synth. Catal.*, 2006, **348**, 813–820.
16. T. Maki, K. Ishihara and H. Yamamoto, *Org. Lett.*, 2006, **8**, 1431–1434.
17. K. Ishihara, T. Maki and H. Yamamoto, *Tetrahedron*, 2007, **63**, 8645–8657.
18. K. Ishihara, S. Ohara and H. Yamamoto, *J. Org. Chem.*, 1996, **61**, 4196–4197.
19. K. Ishihara, S. Kondo and H. Yamamoto, *Synlett*, 2001, **9**, 1371–1374.
20. T. Maki, K. Ishihara and H. Yamamoto, *Org. Lett.*, 2005, **7**, 5043–5046.
21. K. Arnold, A. S. Batsanov, B. Davies and A. Whiting, *Green. Chem.*, 2008, **10**, 124–134.
22. S. Liu, Y. Yang, X. Liu, F. K. Ferdousi, A. S. Batsanov and A. Whiting, *Eur. J. Org. Chem.*, 2013, **25**, 5692–5700.
23. R. M. Al-Zoubi, O. Marion and D. G. Hall, *Angew. Chem., Int. Ed.*, 2008, **47**, 2876–2879.
24. N. Gernigon, R. M. Al-Zoubi and D. G. Hall, *J. Org. Chem.*, 2012, **77**, 8386–8400.
25. J. E. Anderson, J. Cobb, R. Davis, P. J. Dunn, R. N. Fitzgerald and A. J. Pettman, *Industrial Applications of Boric acid Boronic Acid-catalyzed Direct Amidation Reactions*, Published by Wiley and Sons, Inc., Hoboken, New Jersey, 2013, pp. 111–120.
26. L. Kleinberg, S. A. Grossman, K. Carson, G. Lesser, A. O'Neill, J. Pearlman, P. Phillips, T. Herman and M. Gerber, *J. Clin. Oncol.*, 2002, **20**, 3149–3155.

CHAPTER 14

Synthetic Biology for Organic Syntheses

BIRGIT WILTSCHI[a] AND ANTON GLIEDER*[a,b]

[a] acib – Austrian Centre of Industrial Biotechnology, Petersgasse 14, A-8010 Graz, Austria; [b] Institute of Molecular Biotechnology, NAWI Graz, Graz University of Technology, Petersgasse 14, A-8010 Graz, Austria
*Email: glieder@glieder.com

14.1 Synthetic Biology: A New Branch of Synthetic Chemistry?

14.1.1 Synthetic Organic Chemistry Goes Green

The history of mankind is inextricably linked with chemistry: prehistoric man using metals; medieval alchemists frantically searching for a way to make gold; or modern synthetic chemists designing synthesis schemes for the next blockbuster compound for almost 200 years; all of them have been driven by the motivation to produce something new and valuable. It is a stunning fact that a highly advanced science such as modern synthetic organic chemistry has roots in initial trial and error approaches that were entirely ignorant of the underlying chemical principles. However, it clearly illustrates that is not necessary to (completely) understand the details of a system in order to manipulate it—a lesson we learned again in the past 20 years when directed evolution brought a step change in biocatalyst development for chemical synthesis.

RSC Green Chemistry No. 46
Green and Sustainable Medicinal Chemistry: Methods, Tools and Strategies for the 21st Century Pharmaceutical Industry
Edited by Louise Summerton, Helen F. Sneddon, Leonie C. Jones and James H. Clark

Published by the Royal Society of Chemistry, www.rsc.org

Chemists had predominantly analyzed and studied chemical molecules before they finally began to synthesize them. When the synthesis of simple organic molecules, such as urea, became feasible at the beginning of the nineteenth century, their chemical synthesis and the theories explaining their structures rapidly started to develop concurrently.[1] Design and understanding mutually stimulated each other and consequently led to the manufacture of new, unprecedented synthetic compounds that did not exist in nature.

However, since the early days of synthetic organic chemistry, the cheap production of high amounts of (valuable) compounds has been a major concern. Though these production processes have been seizing profitable business opportunities, their frequently adverse impact on human health and the environment has moved into focus more recently. Many of the traditional organochemical synthesis processes involve substances, *e.g.* aggressive or explosive liquids or gases, expensive metal catalysts and organic solvents, that are toxic to humans or are environmentally malign, specifically if applied on a large scale. In other cases, the energy efficiency of chemical processing may be low because extreme pressures and temperatures are employed. Many processes rely on depleting feedstocks, including raw materials that are mined or generated from fossil fuels (petroleum, natural gas or coal). Yet other chemical transformations produce large amounts of hazardous waste, which is a major challenge in chemistry for pharmaceutical production. In contrast to petrochemistry, which developed to be an almost zero waste industry, the different processes, routes and technologies in the pharma industry produce relatively small amounts of high-value products in comparison to large amounts of (sometimes hazardous) waste. The term 'green chemistry' was coined to address exactly these shortcomings: It describes pollution prevention approaches that aim at nontoxic, environmentally benign, and sustainable technological solutions in the design, manufacture, and application of chemical products.[2]

14.1.2 Biochemistry is Green *per se* in Many Aspects

Man has long exploited the chemistry of living organisms, known as biological chemistry or else biochemistry. We have been biotechnologists through the past centuries as much as we have been chemists: for instance, beer brewing, wine making and baking bread are ancient technologies that have accompanied mankind from its early dawn until the present day. Even though our ancestors were not aware of this fact, they cultivated microorganisms and exploited the talents of these chemists from nature. Microorganisms and other cells are very talented natural chemists indeed: they are capable of producing complex compounds, such as alkaloids, antibiotics or terpenoids, from simple carbon sources, such as CO_2, carbon monoxide, energy, hydrogen, oxygen, water and salts.

However, although the product palette available from naturally occurring organisms is broad, it is still predefined by evolution as they adapted their biochemistries for survival in specific biological niches. For instance, the mold *Penicillium* produces the β-lactam antibiotic penicillin and secretes it into its environment in order to suppress bacterial competitors for nutrients. We exploit the same antibiotic, which efficiently secures the survival of the fungus, for the treatment of infectious diseases caused by bacteria. The list of useful natural products from microorganisms has become ever longer since the discovery of the antibiotic activity of penicillin. Sophisticated cultivation, mutation and selection techniques were developed to elevate the titers of the valuable substances in the natural production hosts. Later, genetic engineering facilitated the targeted manipulation of the genetic information encoding the biosynthesis pathways. Moreover, the production hosts could be equipped with the genetic information for the biosynthesis of natural compounds that they originally were not capable of producing,[3] *e.g.*, the production of the human peptide hormone insulin in the bacterium *E. coli*.[4]

Living organisms usually do not occur as isolated entities; rather they are part of an ecosystem. To persist, ecosystems must be self-sustainable and the recycling of waste is a central issue. Furthermore, biological chemistry has evolved in and for aqueous systems as water is the most abundant solvent on our planet. Biosystems use biological macromolecules, such as peptides and polynucleotides, as biocatalysts, which are powerful accelerators of chemical reactions at ambient pressure and temperature. This modular way of construction from a few relatively simple universal building blocks to long polymer chains provides an unlimited diversity of catalyst structures made by a standardized machinery of life for their manufacture. Briefly, living organisms and specifically microorganisms appear to offer sustainable chemistries that perfectly match the concepts of green chemistry and the concepts of industrial engineering.

14.1.3 Synthetic Biology: Design and Synthesis of Biological Systems

Modern biology is in a very similar situation to how chemistry was at the beginning of the nineteenth century. While classical biology is devoted to studying life, we have recently learned how to manipulate the genetic information of biological systems. Consequently, we are currently on the verge of taking the next step, which is the deliberate design of biological systems. Even if we do not understand them to the minutest detail, our present knowledge suffices to customize them for useful (biosynthetic) purposes in order to get access to new products and products of higher quality. In turn, biotechnological production will increase our knowledge about biosystems. We are about to transfer to biology the principles that previously helped organic synthetic chemistry gain the importance it has for our modern lives: design and synthesis by understanding.

One ambitious aim of the emerging scientific field of synthetic biology is to deliver biological systems for the next generation of deliberate synthetic chemistry.

14.2 The Emerging Discipline of Synthetic Biology

14.2.1 What Is Synthetic Biology?

There has been a big debate in the life sciences community about what exactly synthetic biology would be and, accordingly, numerous definitions were presented, *e.g.* see www.syntheticbiology.org. All definitions including a recent definition published by the European ERA-NET Synthetic Biology consortium (www.erasynbio.eu) agree that synthetic biology is about applying the principles of engineering to biology for useful purposes: Using synthetic biology tools, biological systems or their parts can be deliberately designed and constructed in order to serve a certain purpose. Thus, in synthetic biology the emphasis is on specific goals and applications in the context of biological systems.[1]

Metabolic reprogramming and the construction of living "production plants" for, *e.g.*, the manufacture of complex chemical compounds from cheap basic precursors is a key topic in synthetic biology.[5] Further research areas include "bio-computing" using synthetic gene networks; synthetic genomics and genome minimization; protocells and the exploration of the origins of life; chemical synthetic biology with non-standard building blocks (DNA bases, amino acids); as well as the biosafety and societal aspects of synthetic biology.

14.2.2 Parts, Devices and Systems: Basic Principles of Synthetic Biology

In comparison to classical biology, synthetic biology adopts a fundamentally different notion of biological systems. So-called chassis cells are construed as living entities that possess a basic metabolism to produce primary metabolites and are capable of essential chemical transformations, such as energy production and cofactor regeneration. The chassis can be furnished with a desired function, *e.g.*, the ability to produce a specific compound. Whether the compound is natural or non-natural does not matter as long as the genetic information driving its biosynthesis is available or can be designed. The cell chassis is equipped with an appropriate genetic system (a gene network) that consists of devices (*e.g.*, different DNA vectors) made of parts (*i.e.*, regulatory DNA sequences such as promoters, terminators and ribosome binding sites that drive the expression of, *e.g.*, genes encoding enzymes; Figure 14.1).

However, even in the simplest unicellular life forms, genetic and metabolic networks never work independently but are intricately involved in complex interactions with their cellular environment. This is why the chassis

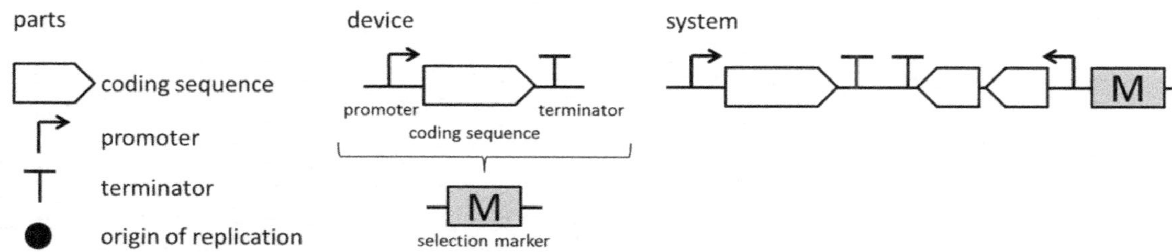

Figure 14.1 Parts, devices and systems. Parts are basic genetic elements, such as a promoter, a terminator, an origin of replication, a coding sequence (*e.g.* for enzymes as parts of pathways) or the like. Parts can be assembled into devices. For instance, a functional transcription unit consists of a promoter, a coding sequence and a terminator. Devices, such as the functional transcription unit for a selection marker, may be specified using appropriate symbols. Systems consist of various devices.

cell concept as outlined above is too simplistic.[6] Nevertheless, the idea to apply a modular approach for the systematic assembly of higher order (genetic) systems from well-defined basic parts is very useful and essential for an automatized approach as it is common in industrial engineering. Accordingly, powerful strategies to link the parts together in an ordered, parallel or combinatorial manner have been developed.[7,8]

14.3 Opportunities Synthetic Biology offers Green Chemistry

14.3.1 Synthetic Biology in Service to Synthetic Chemistry?

Synthetic biology constitutes the next development step ensuing from biotransformations and metabolic engineering. Classical biotransformations[9] use an enzymatic biocatalyst for the chemical transformation of a specific substrate (or substrate class). Enzyme cascade reactions expand this principle towards more complicated chemical transformations as they employ a series of consecutive enzyme catalysts. Similarly, multi-enzyme one-pot reactions allow for example redox cofactor regeneration in NAD(P)H dependent redox reactions. From a very simplistic perspective, a cellular biosynthesis pathway may be regarded as a (long) enzyme cascade that is plugged into the host cell metabolism to be supplied with substrate(s) and cellular regeneration of cofactors. Along the pathway, chemical intermediates are produced and used for the next reaction step without the need to purify them in between. Metabolic engineering facilitates the manipulation of the intracellular metabolic reactions such that the output of the desired natural product is maximized.[10] For instance, heterologous genes are overexpressed to facilitate the (over)production of desired natural metabolites. Otherwise, the intracellular flux of substrates and precursors can be redirected towards product synthesis. Synthetic biology extends this approach to non-natural pathways and non-natural products, and frequently applies new tools of modern molecular biology and gene synthesis. Using parts and devices from different sources, *i.e.* different kingdoms of life, biosynthesis pathways can be constructed that do not exist as such in extant life forms but catalyze the production of the desired compound. DNA synthesis allows the removal of unknown and hardly controllable natural regulation mechanisms to get predictable and simple as well as scalable control over the new pathways in a non-natural industrial environment. Combinatorial pathway assembly methods facilitate the application of entire enzyme toolboxes for biocatalytic reactions instead of only one enzyme at a time. This approach allows for but also relies on efficient screening to identify the most productive pathway.

As the cell provides a protecting environment for the enzymes, their choice is not confined to the simplest and most stable biocatalysts. Biocatalysts that turn out to be labile when used in isolated form in biotransformation reactions, such as some large enzyme complexes or membrane-bound enzymes, can be well included in synthetic pathway design. Moreover, the host cells are equipped with effective regeneration systems for cofactors, such as NADH (nicotinamide adenine dinucleotide), NADPH (nicotinamide adenine dinucleotide phosphate) and ATP (adenosine triphosphate). However, if the natural efficacy should still be insufficient for a specific application, synthetic biology offers genomic engineering tools to optimize the regeneration pathways of the chassis. Often, metabolic models derived from systems biology studies support their design.

14.3.2 Opportunities for Synthetic Biology in Green Chemistry

Synthetic biology has something to offer to synthetic chemists. Exploiting biosynthesis instead of specific organic chemistries, either versatile chemical intermediates and building blocks or useful end-products are accessible. These compounds are either produced *de novo* using a simple, inexpensive carbon source, salts and an energy source, or chemically synthesized precursors may be refined or functionalized by regio-, chemo- or stereo-selective chemical modifications, such as hydroxylation, alkylation, or selective oxidations or halogenations.

14.3.3 A Synthetic Biology Approach for the Biosynthesis of Artemisinin

One of the most cited paradigms for applied synthetic biology is the biosynthesis of the artemisinin precursor artemisinic acid in the baker's yeast *S. cerevisiae*. The antimalarial drug artemisinin is a sesquiterpene endoperoxide naturally produced by the plant *Artemisia annua* (sweet wormwood). Jay Keasling and colleagues developed an efficient biosynthetic route to the precursor artemisinic acid that is coupled to a practical, efficient and scalable chemical process for the conversion of artemisinic acid to artemisinin.[11] This successful combination of a biosynthetic with a synthetic route very well illustrates how organic synthetic chemistry and synthetic biology amalgamate towards efficient industrial production processes.

The non-natural pathway exploits the yeast metabolic network for farnesyl diphosphate (FPP) synthesis onto which the artemisinic acid biosynthesis genes from *A. annua* are plugged (Figure 14.2).

To provide sufficient amounts of FPP, the yeast pathway starting from acetyl-CoA is up-regulated and the pathway flux through the downstream reactions from FPP to the end-product ergosterol is restricted by repressing the yeast squalene synthase (Erg9). This leads to a substantial accumulation of FPP, which is efficiently converted to amorpha-9,11-diene by the recombinantly overexpressed amorphadiene synthase (ADS) from *A. annua*. A three-step oxidation reaction transforms amorpha-9,11-diene into artemisinic acid. All steps of the first functional artemisinic acid pathway in yeast are catalyzed by *A. annua* enzymes: The *A. annua* cytochrome P450 amorphadiene oxidase CYP71AV1 and its cognate reductase CPR1 from *A. annua* hydroxylate amorpha-9,11-diene to artemisinic alcohol; co-expression of the cytochrome b5 CYB5 from *A. annua* greatly improves this hydroxylation step. Alcohol dehydrogenase (ADH1) oxidizes the primary alcohol to artemisinic aldehyde and, finally, artemisinic aldehyde dehydrogenase

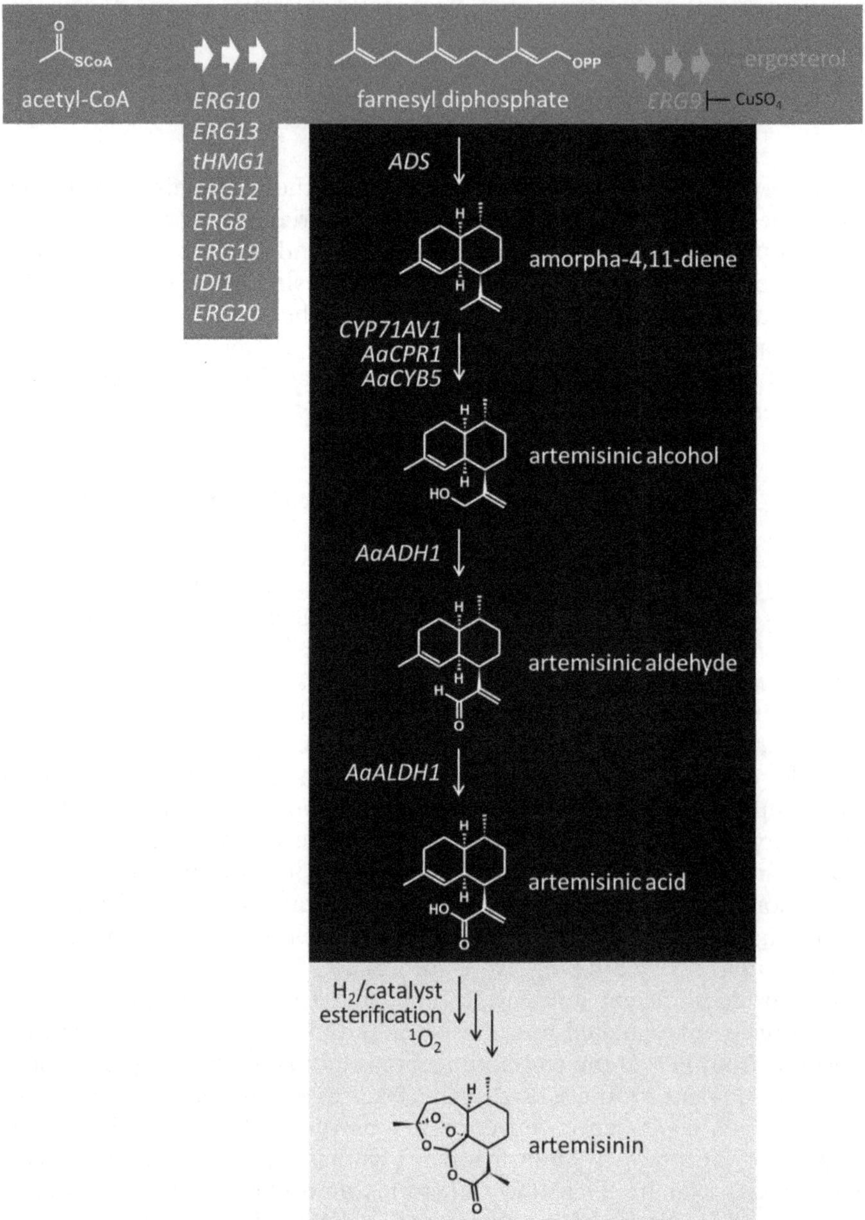

(ALDH1) completes the formation of artemisinic acid from the aldehyde precursor by another oxidation reaction. In order to balance the flux through the non-natural pathway, some biosynthesis genes are expressed from high copy plasmids, others are integrated into the yeast genome and all genes are controlled by inducible promoters of varying strength (Figure 14.3).

Up to 25 g L^{-1} artemisinic acid can be produced by the described approach.[11] Since artemisinic acid is extracted from the fermentation broth and then chemically converted, the final product and active pharmaceutical ingredient is not a direct product from microbial fermentation.

14.3.4 Vanillin Production in Yeasts

The non-natural biosynthesis of vanillin in the yeasts *Schizosaccharomyces pombe* and *Saccharomyces cerevisiae* is an example for the application of synthetic biology for the full *de novo* synthesis of a valuable end-product.[12] Vanillin (3-methoxy-4-hydroxybenzaldehyde) is the main component of vanilla flavor. It is the global leader in aroma compounds and most of the vanillin sold annually originates from chemical synthesis from lignin or fossil hydrocarbons, in particular from the naturally occurring aromatic oil guaiacol. Hansen *et al.* devised an alternative approach[12] similar to that for artemisinin biosynthesis developed by Keasling: In addition to fission yeast, they also used the generally recognized as safe (GRAS) baker's yeast as one of the hosts for the new biosynthetic pathway. In addition, they extended an endogenous pathway of the yeast's primary metabolism with the appropriate reaction steps leading to the desired end-product.

Dehydroshikimic acid from the aromatic amino acid biosynthesis pathway of yeasts serves as a precursor for the production of the vanillin (Figure 14.4).

Figure 14.2 Overview of artemisinin production by synthetic biology.[11] In order to increase the flux through the yeast mevalonate pathway, all genes (*ERG10*, *ERG13*, *tHMG1*, *ERG12*, *ERG8*, *ERG19*, *IDI1*, *ERG20*) for the conversion of acetyl-CoA to farnesyl diphosphate (FPP) were overexpressed. To restrict formation of the end-product ergosterol and provoke accumulation of the pathway intermediate FPP, the gene encoding squalene synthase (*ERG9*) is under transcriptional control of a $CuSO_4$-respressible promoter. FPP is cyclized by *Artemisia annua* amorphadiene synthase (ADS) to form amorpha-4,11-diene, which is oxidized to artemisinic acid in three successive steps. In the first step, the *A. annua* cytochrome P450 amorphadiene oxidase CYP71AV1 together with its cognate reductase AaCPR1 and the auxiliary cytochrome b5 AaCYB5 hydroxylate amorphadiene. The *A. annua* alcohol dehydrogenase AaADH1 oxidizes the artemisinic alcohol to its aldehyde, which is finally oxidized to artemisinic acid by artemisinic aldehyde dehydrogenase (AaALDH1). Artemisinic acid is extracted from the culture medium and converted to artemisinin by chemical synthesis using chemically generated singlet oxygen. Dark grey box, *S. cerevisiae* genes; black box, *A. annua* genes; light grey box, chemical processing of artemisinic acid to artemisinin.

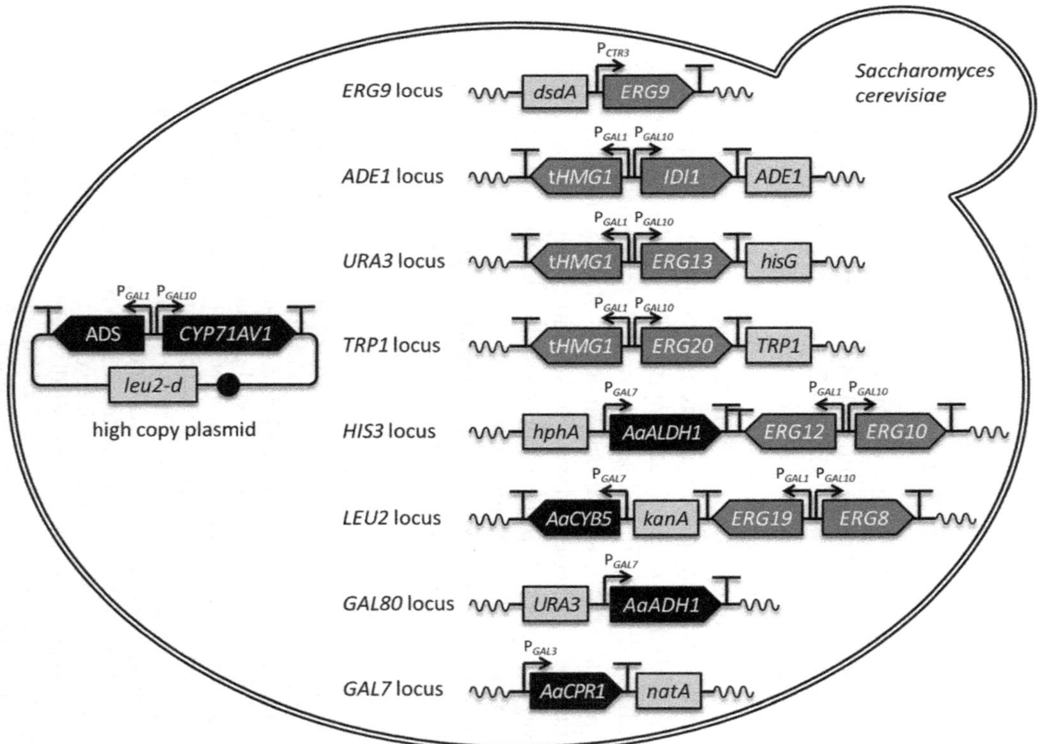

Figure 14.3 Metabolic reprogramming of *S. cerevisiae* for the biosynthesis of artemisinic acid.[11] The genes encoding the yeast mevalonate pathway enzymes (dark grey) were expressed from the strong, divergent *GAL1/10* promoter. Three copies of t*HMG1* were inserted at three different genomic loci for the efficient overproduction of HMG-CoA reductase. The canonical *ERG9* promoter was replaced by a $CuSO_4$-repressible *CTR3* promoter. Thus, expression of *ERG9* is repressed in the presence of $CuSO_4$. The *Artemisia annua* genes (black) *AaADH1*, *AaALDH1*, *AaCYB5* and *AaCPR1* were expressed using the strong *GAL7* or *GAL3* promoters, respectively. The expression cassettes for the yeast mevalonate pathway enzymes and the *A. annua* genes were integrated into the yeast genome at 8 different genomic loci. In contrast, the *A. annua ADS* and *CYP71AV1* genes were expressed from a high copy plasmid under the transcriptional control of the divergent *GAL1/10* promoter. Light grey boxes indicate selection marker devices; wavy lines symbolize the genome environment at the insertion locus. For an explanation of the parts symbols refer to Figure 14.1.

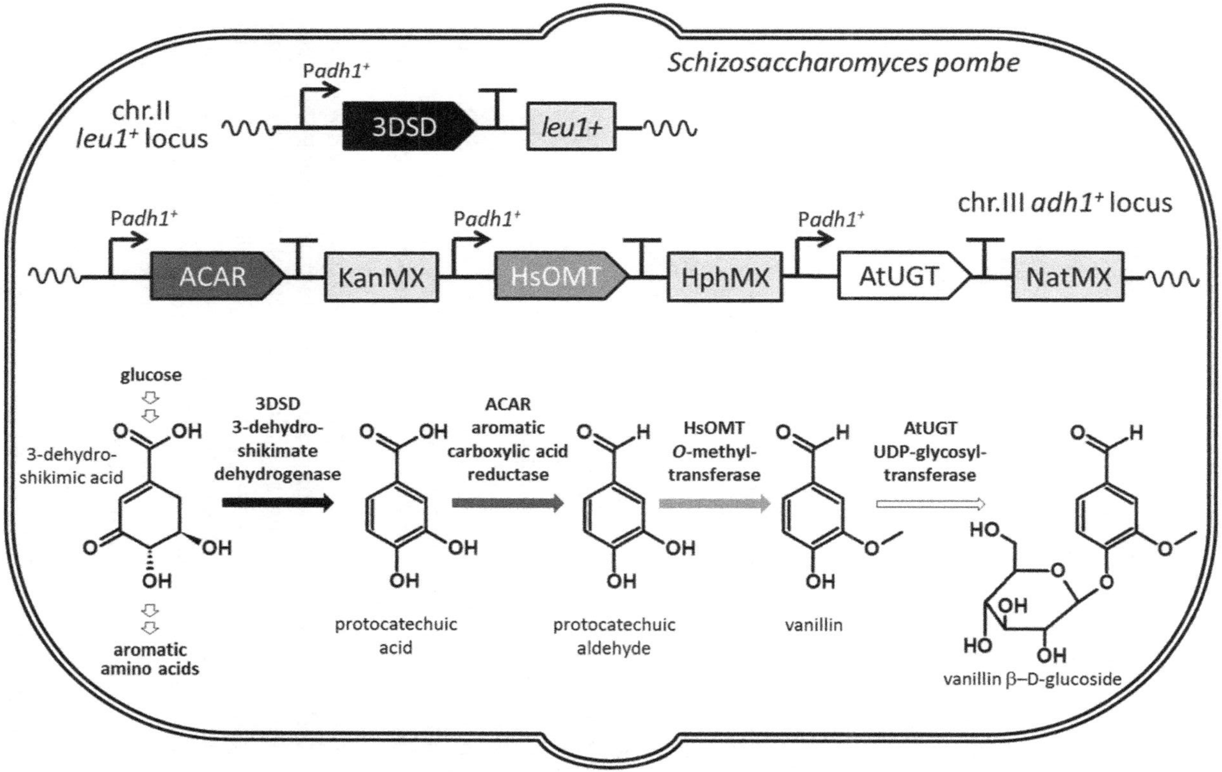

Figure 14.4 Biosynthesis of vanillin and vanillin beta-D glucoside in *S. pombe*. Two expression cassettes for the heterologous genes were generated. One cassette contained only the coding gene for 3DSD (black) while the other cassette carried functional transcription units for ACAR (dark grey), Hs-OMT (grey) and AtUGT (white). The cassettes were introduced into the *S. pombe* genome at two different loci. Light grey boxes indicate selection marker devices; wavy lines symbolize the genome environment at the insertion locus. For an explanation of the parts symbols refer to Figure 14.1.

This vanillin precursor is converted to vanillin by three subsequent enzymatic steps: first, the 3-dehydroshikimate dehydratase (3DSD) from the dung mold *Podospora pauciseta* catalyzes the conversion of 3-dehydroshikimic acid to protocatechuic acid. The next step, the ATP-driven reduction of protocatechuic acid to protocatechuic aldehyde, is catalyzed by the aromatic carboxylic acid reductase (ACAR; EC 1.2.1.30) from *Nocardia* spp. Finally, the human *O*-methyltransferase (OMT) methylates the *meta*-hydroxyl group of protocatechuic aldehyde to yield vanillin. Furthermore, OMT is able to methylate protocatechuic acid to yield vanillic acid, which in turn is reduced to vanillin by ACAR. Hansen *et al.* integrated the genes encoding 3DSD, ACAR and OMT into the genome of *S. pombe* and *S. cerevisiae*, and expression was driven by the strong constitutive *adh1* + and *TPI1* promoters, respectively. In *S. cerevisiae*, the ACAR enzyme required activation by phosphopantetheinylation, which was achieved by coexpression of a *Corynebacterium glutamicum* phosphopantetheinyl transferase (PPTcg-1) from a low copy (*CEN-ARS*) plasmid.

As vanillin is toxic to the yeast cells, it was converted into vanillin β-D-glucoside using *Arabidopsis thaliana* UDPglucose glycosyltransferase. Vanillin β-D-glucoside is more water soluble and less cytotoxic than vanillin, which is why the glycosylation alleviates the inhibitory effect of free vanillin and facilitates its accumulation in higher amounts. 65 mg L^{-1} and 45 mg L^{-1} vanillin were produced in *S. pombe* and *S. cerevisiae*, respectively, which is quite satisfactory for a non-optimized process.

The company Evolva successfully completed the development of vanillin during 2014 and its partner International Flavors & Fragrances Inc. (IFF) has been commercializing the product since mid-2014 (http://www.evolva.com/products/vanillin).

Besides artemisinin and vanillin, other high-value compounds have already been produced, mostly from fermentable sugars, by applying the tools of synthetic biology, *e.g.*, Ω-3 fatty acids and the anti-cancer agent taxol;[13] hydrocortisone (Sanofi);[14] alkaloids;[15] resveratrol (Evolva); fuel and chemical feedstocks, such as microbial 1,3-propanediol (DuPont Tate & Lyle) and isobutanol (GEVO).[16] However, applications of synthetic biology are not necessarily limited to converting sugar to commercially interesting compounds, they are also increasingly appreciated as complementary steps in classical synthetic routes: They can be used to provide chemical building blocks and precursors from renewable carbon sources, to replace individual steps of synthetic routes, to create short cuts or to extend the chemical and functional diversity of the final products.

14.4 Limitations of Synthetic Biology for Green Chemistry

Though synthetic biology has something to offer to synthetic chemists, it is not applicable for every kind of compound. Moreover, not every bioprocess enabling the production of valuable compounds from renewable resources, such as fermentable sugars, is necessarily "greener" than chemical synthesis from exhaustible feedstock, *e.g.* fossil oil. On the one hand, natural resources such as polysaccharides are highly oxidized and activated compounds. Thus, it takes many chemical transactions to reduce them to typical building blocks for organic synthesis. On the other hand, product titers in fermentations are often low and downstream processing causes a strong economic and ecologic impact. Though synthetic biology expands the scope of biotransformations and metabolic engineering, the products that can be produced in a microbial host are still limited to those that are accessible by today's enzymatic toolboxes of biocatalysis. If biocatalysis is inefficient or confined to particular substrates or products, as, *e.g.*, with Michael additions[17] or fluorination,[18] organosynthesis indeed is the preferred choice.

Today's synthetic biology approaches are also time consuming and typical success stories from the past, such as the production of 1,3-propanediol and artemisinic acid, needed more than 100 person years each to obtain a first reasonable amount of product. This is expected to change by standardization of parts, the availability of such parts *via* registries and increasing efficiency and capacities in gene and genome synthesis, as well as lab automation.

Due to the novelty of the synthetic approaches and applied technologies, the public and even the regulatory authorities might also regard chemical compounds produced by applied synthetic biology as new and therefore less safe than the same compounds derived from an established organosynthetic route. Consequently, synthetic biology products are very critically assessed, no matter whether they are chemically identical to their organosynthetic counterparts. This critical assessment and detailed product analyses are facilitated by the fact that synthetic biology employs a limited number of usually safe host chassis. These provide a safe and very well-known biological matrix as compared to natural extracts from plants, fungi and other organisms. In other cases, the fermentation products run through a number of final chemical processing and purification steps before their final use.

Despite potential flaws, synthetic biology has great potential for green chemistry. High value products, such as artemisinic acid and vanillin, can now be reliably produced in chemically pure form and under strictly defined, reproducible conditions, employing preferred production hosts, such as *S. cerevisiae*, *i.e.* microorganisms that are *g*enerally *r*egarded *a*s *s*afe (GRAS organisms). If high-value products, such as artemisinin or vanillin, are isolated from their natural plant sources, their yield and composition is dependent on many different, hard to control parameters, such as weather, rainfall or pests. Moreover, the land footprint of active ingredients extracted from plants is typically significantly larger than for fermentation products and plant extracts often contain mixtures of the desired compound with isoforms or contaminants.

14.5 Summary

- Synthetic biology constitutes the next development step ensuing from biotransformations and metabolic engineering.
- Chassis are equipped with parts and devices to make useful compounds with a lower land footprint than, for example, secondary metabolites extracted from plants.
- Living cells are very talented natural chemists producing complex compounds such as alkaloids, antibiotics or terpenoids from simple carbon sources such as CO_2 and carbon monoxide, energy, hydrogen, oxygen, water and salts.
- Synthetic biology offers complementary synthetic routes but is not necessarily always greener than chemical synthesis from exhaustible carbon sources.
- Synthetic biology processes are scalable and secure a reliable supply of natural building blocks.

Acknowledgements

This work has been supported by the Federal Ministry of Science, Research and Economy (BMWFW), the Federal Ministry of Traffic, Innovation and Technology (bmvit), the Styrian Business Promotion Agency SFG, the Standortagentur Tirol, the Government of Lower Austria and ZIT - Technology Agency of the City of Vienna through the COMET-Funding Program managed by the Austrian Research Promotion Agency FFG (grant number 282482).

Support by CHEM21 in the frame of the Innovative Medicines Initiative Joint Undertaking under grant agreement no 115360, resources of which are composed of financial contribution from the European Union's Seventh Framework Programme (FP7/2007-2013) and EFPIA companies' in-kind contribution is acknowledged.

References

1. B. J. Yeh and W. A. Lim, *Nat. Chem. Biol.*, 2007, **3**, 521.
2. P. T. Anastas and J. C. Warner, *Green Chemistry: Theory and Practice*, Oxford University Press, New York, 1998, p. 150.
3. J. D. Keasling, *Metab. Eng.*, 2012, **14**, 189.
4. D. C. Williams, R. M. Van Frank, W. L. Muth and J. P. Burnett, *Science*, 1982, **215**, 687.
5. V. G. Yadav, M. De Mey, C. Giaw Lim, P. Kumaran Ajikumar and G. Stephanopoulos, *Metab. Eng.*, 2012, **14**, 233.
6. A. Danchin, *FEBS Lett.*, 2012, **586**, 2129.
7. T. Ellis, T. Adie and G. S. Baldwin, *Integr. Biol. (Camb.)*, 2011, **3**, 109.
8. T. Wang, X. Ma, H. Zhu, A. Li, G. Du and J. Chen, *Appl. Microbiol. Biotechnol.*, 2012, **93**, 1853.

9. K. Faber, *Biotransformations in Organic Chemistry: A Textbook*, Springer-Verlag, Berlin Heidelberg, 6th revised, and corrected edn., 2011, p. 500.
10. J. D. Keasling, *ACS Chem. Biol.*, 2008, **3**, 64.
11. C. J. Paddon, P. J. Westfall, D. J. Pitera, K. Benjamin, K. Fisher, D. McPhee, M. D. Leavell, A. Tai, A. Main, D. Eng, D. R. Polichuk, K. H. Teoh, D. W. Reed, T. Treynor, J. Lenihan, H. Jiang, M. Fleck, S. Bajad, G. Dang, D. Dengrove, D. Diola, G. Dorin, K. W. Ellens, S. Fickes, J. Galazzo, S. P. Gaucher, T. Geistlinger, R. Henry, M. Hepp, T. Horning, T. Iqbal, L. Kizer, B. Lieu, D. Melis, N. Moss, R. Regentin, S. Secrest, H. Tsuruta, R. Vazquez, L. F. Westblade, L. Xu, M. Yu, Y. Zhang, L. Zhao, J. Lievense, P. S. Covello, J. D. Keasling, K. K. Reiling, N. S. Renninger and J. D. Newman, *Nature*, 2013, **496**, 528.
12. E. H. Hansen, B. L. Møller, G. R. Kock, C. M. Bünner, C. Kristensen, O. R. Jensen, F. T. Okkels, C. E. Olsen, M. S. Motawia and J. Hansen, *Appl. Environ. Microbiol.*, 2009, **75**, 2765.
13. V. M. Ye and S. K. Bhatia, *Biotechnol. J*, 2012, **7**, 20.
14. F. M. Szczebara, C. Chandelier, C. Villeret, A. Masurel, S. Bourot, C. Duport, S. Blanchard, A. Groisillier, E. Testet, P. Costaglioli, G. Cauet, E. Degryse, D. Balbuena, J. Winter, T. Achstetter, R. Spagnoli, D. Pompon and B. Dumas, *Nat. Biotechnol.*, 2003, **21**, 143.
15. K. M. Hawkins and C. D. Smolke, *Nat. Chem. Biol.*, 2008, **4**, 564.
16. E. I. Lan and J. C. Liao, *Bioresour. Technol.*, 2013, **135**, 339.
17. E. Zandvoort, E. M. Geertsema, B.-J. Baas, W. J. Quax and G. J. Poelarends, *Angew. Chem., Int. Engl.*, 2012, **51**, 1240.
18. C. Schaffrath, H. Deng and D. O'Hagan, *FEBS Lett.*, 2003, **547**, 111.

CHAPTER 15

Biocatalysis for Medicinal Chemistry

A. S. WELLS

Charnwood Technical Consulting Ltd, Parklands, Northage Close, Quorn, Leicestershire LE12 8AT, UK
Email: charnwoodtechnicalconsulting@gmail.com

15.1 Introduction to Recent Advances in Biocatalysis

Over the past 5 years or so, there has been an upsurge of interest from medicinal chemistry groups in embracing the philosophy and tools of green chemistry.[1] This philosophy is in part a driver to move towards more sustainable practices, but there is also interest in using emerging new technologies to speed up the drug discovery process and to discover new and diverse structures as scaffolds and lead compounds. Biocatalysis and industrial biotechnology can offer a range of effective and sustainable green solutions across the spectrum of pharmaceutical R&D, manufacturing and life cycle management (Figure 15.1).

In the discovery arena, biocatalysis can offer the following:

- ➢ Rapid synthesis of metabolites for adsorption, distribution, metabolism and excretion/drug metabolism and pharmacokinetic (ADME/DMPK) studies.
- ➢ Unique catalytic opportunities—either not accessible with conventional chemistry, or allowing efficient shorter routes to complex chiral intermediates.

RSC Green Chemistry No. 46
Green and Sustainable Medicinal Chemistry: Methods, Tools and Strategies for the 21st Century Pharmaceutical Industry
Edited by Louise Summerton, Helen F. Sneddon, Leonie C. Jones and James H. Clark

Published by the Royal Society of Chemistry, www.rsc.org

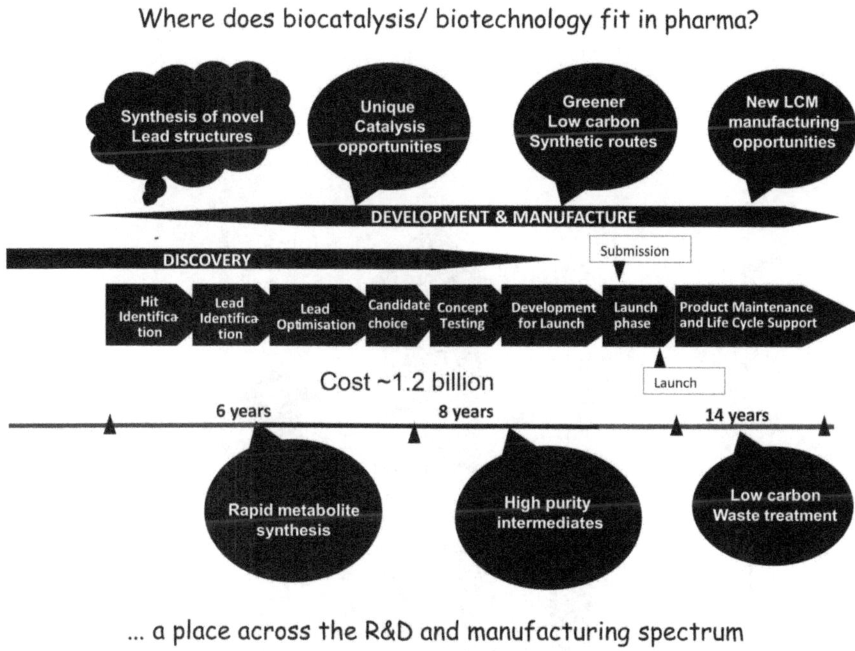

Figure 15.1 Biocatalysis/biotechnology in the pharmaceutical industry.

- ➢ Access to a wider range of novel lead structures for drug discovery programs.
- ➢ The synthesis of chiral compounds in high enantiomeric excess (ee) has been, and will be in the future, a key component of medicinal chemistry.[2] Biocatalysis is increasingly a pivotal technology in the tool box of chiral synthesis.

Looking at the history of biocatalysis as a tool for synthetic organic chemists, many of the barriers in obtaining sufficient amounts of active protein, a sufficiently wide diversity of biocatalysts, and enzymes that have robust performance have now been solved.[3] Many of the advanced tools used to study and evolve proteins can be applied to the development of biocatalysts, making targeted advances in enzyme performance in months that would have been impossible or too impractical even a few years ago.[4] An example is shown in Figure 15.2.

In the late 1970s and early 1980s reports of 2-deoxy-D-ribose 5-phosphate aldolase (DERA) enzymes as biocatalysts to make highly functionalised chiral synthons began to appear in the organic chemistry literature. However, it was not until early 2000s that DSM solved the issues in getting this enzyme to work under conditions that would make it an attractive industrial catalyst.[5] Now DERA enzyme mutants are used on large (multi-tonne) scale to make synthons that can be converted into the chiral 1,3-diols used in the manufacture of the HMG-CoA reductase inhibitors rosuvastatin (Crestor®)

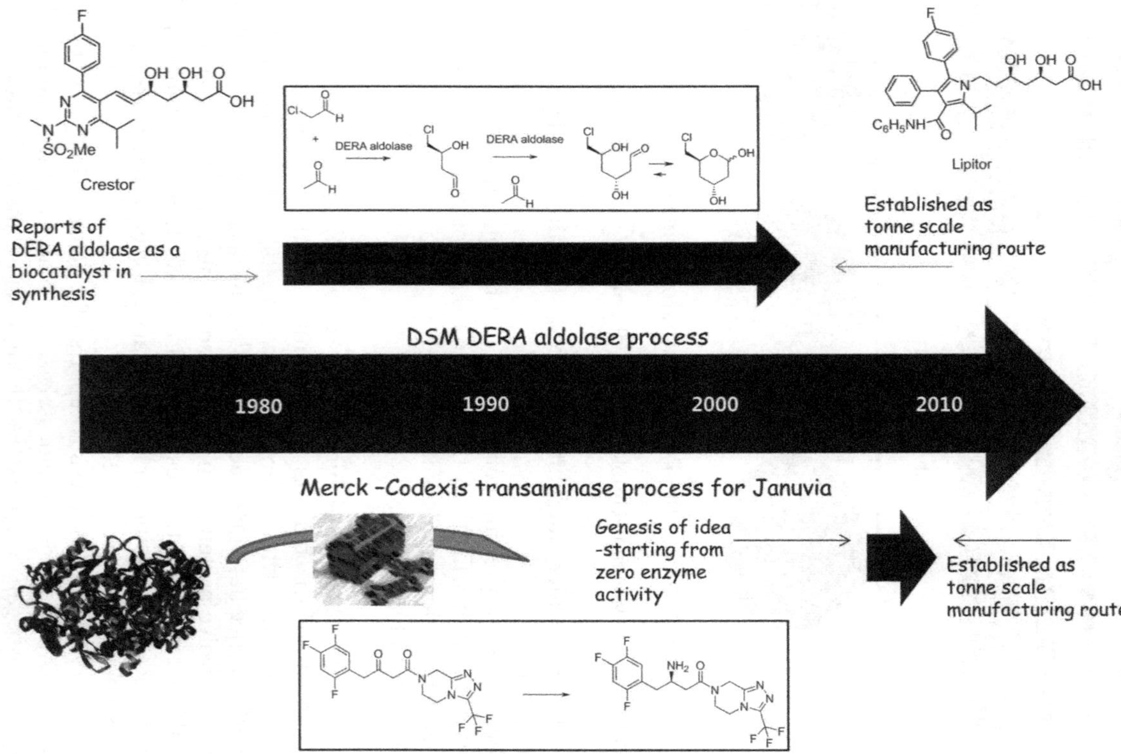

Figure 15.2 Historical and current timelines for the development of biocatalysts for use in organic synthesis.

and atorvastatin (Lipitor®). Contrast this time line to the Merck–Codexis development of a novel ω-transaminase used for the production of the DDP-4 inhibitor sitagliptin (Januvia®), used to treat diabetes mellitus.

In a program utilizing all the tools of modern molecular biology, the Merck–Codexis team developed a novel enzyme that was the basis for a greener industrial process in ~18 months—starting from zero enzyme activity with the desired substrate.[6] This was achieved by evolving a series of transaminase enzymes to accept increasingly larger and more complex substrates until the desired activity with the sitagliptin ketone was achieved. The project won Merck–Codexis a USA Presidential Green Chemistry Challenge award in 2010.

The ability to rapidly produce an efficient bespoke biocatalyst for a chemical process has been driven by advances in techniques like bioinformatics, facile access to protein structures, dramatic lowering of cost and time for sequencing and gene synthesis, high-throughput screening, analytics and a variety of techniques to evolve libraries of mutant enzymes.[7] This has led to an increase in both the number of enzyme-catalysed processes and an increased range of enzyme types being used in the synthesis of pharmaceutical intermediates.[8] The widespread use of heterologous expression has enabled the reliable and cost-efficient supply of many enzymes now routinely used in organic synthesis.

Driven by the major advances in enzyme discovery and optimization, for many process chemistry groups, biocatalysis is quickly becoming a synthetic technique of first choice for certain transformations. It is becoming apparent that biocatalysis is optimally employed when opportunities for enzyme-catalysed transformations are considered at the design stage of synthesis, rather than retro-fitted to solve problems with classical chemical transformations.[9]

Whilst biocatalysts may offer many advantages over conventional synthetic organic chemistry, it should also be borne in mind that significant lifecycle burdens can be generated during the downstream processes for bio-reactions.[10] The product streams from bio-reactions are typically dilute aqueous solutions, which can bring specific problems that need to be addressed from both an environmental and process optimisation standpoint. Any new methodologies developed need to be assessed from a holistic point of view, to ensure potential environmental hotspots are not ignored downstream from the actual synthetic step. This is only a consideration when operating at scale, and not really applicable at the medicinal chemistry scale of operation.

15.2 Some Enzyme Classes Readily Accessible to Medicinal Chemists

In the following section, a number of enzyme classes will be examined. These enzymes can be readily obtained from commercial vendors and used by any competent organic chemist. These biocatalysts can provide rapid and green synthesis of a wide range of structures. For a detailed contemporary review of these enzymes and others useful in synthesis see Volume 7: Synthetic Methods VI—Enzymatic and Semi-Enzymatic and Volume 9: Industrial Applications of Asymmetric Synthesis of Comprehensive Chirality.[11]

15.2.1 Hydrolase Enzymes

Often in the early stages of a drug development project, both enantiomers of a target racemic molecule are needed for biological evaluation. Hydrolytic enzymes (lipases, esterases and proteases) can be employed to resolve enantiomeric mixtures *via* hydrolysis of esters and amides, or alcohols and amines by acylation in organic solvents. Often thought of as a mature class of biocatalysts, new recombinant esterases and proteases continue to be developed. Increasingly, hydrolases are used with meso compounds to produce single enantiomer products in high ee and yield, or in combination with a second chemo- or bio-catalyst to racemise the undesired enantiomer *in situ* to set up a dynamic kinetic resolution.[12] Figure 15.3 highlights some chiral APIs and intermediates that can be prepared using hydrolytic enzymes.

Hydrolytic enzymes are traditionally associated with the synthesis of chiral compounds, but increasingly are being employed for other chemistries, such as selective hydrolysis of positional isomers and separation of mixtures other than racemates. The high regio and functional group selectivity offered by enzymes, and the mild reaction conditions employed, make hydrolase catalysis ideal for the manipulation of complex, highly functionalised molecules. Some examples are shown in Figure 15.4.

15.2.2 Ketone Reductases

Chiral alcohols are very useful molecules in their own right, but also as substrates for further conversion to other synthons like sulfides *etc*. The use of ketone reductase (alcohol dehydrogenase) enzymes has now developed to a point of maturity where this technology has become one of the first choices for several pharma companies looking for green manufacturing of chiral alcohols *via* enantioselective reduction of ketones.[13] A wide range of chiral alcohols can be accessed, and both *S* and *R* selective enzymes have been developed. A selection of the range of chiral alcohols accessible *via* enzymatic ketone reduction is shown in Figure 15.5. A big attraction of biocatalysis is the ability to produce small chiral molecules in high ee—a task often difficult to achieve efficiently and cleanly with traditional chemical catalysts. Ketone reductases work by delivering a hydride to one of the prochiral faces of the ketone substrate. The hydride comes from a co-factor, nicotinamide adenine dinucleotide (NADH) or nicotinamide adenine dinucleotide phosphate (NADPH). To achieve an efficient chemical process, the resulting oxidised co-factors NAD^+/$NADP^+$ need to be recycled. Co-factor recycling has

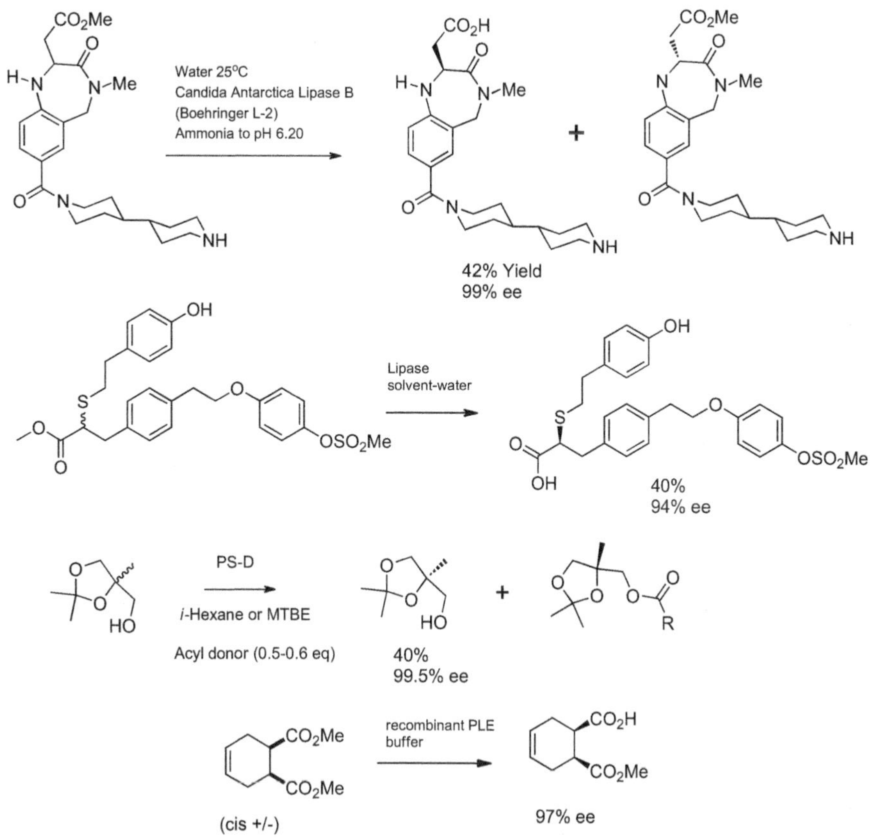

Figure 15.3 Examples of chiral APIs and intermediates accessible *via* hydrolase enzymes.

been the topic of a great deal of recent research, and a range of very efficient chemical and biocatalytic options are now available to recycle NAD(P)H.[14]

15.2.3 ω-Transaminase Enzymes

Alongside the increase in the number of ketone reductases available, a similar situation is found with ω-transaminases. Over the past few years, there has been a rapid rise in the number of these enzymes that can be obtained commercially from a number of suppliers—both *S* and *R* selective enzymes are available. There is also now a great deal of process understanding on getting productive reactions with this enzyme class. ω-Transaminases take an amine donor and a ketone substrate and convert into an amine product and ketone by-product, so driving the equilibrium in the desired direction is essential. Nitrogen-centered chemistry is central to pharmaceutical synthesis and ω-transaminases, along with other enzyme classes, can be used to prepare a wide range of chiral amines.[15] The attraction of the ω-transaminase enzyme

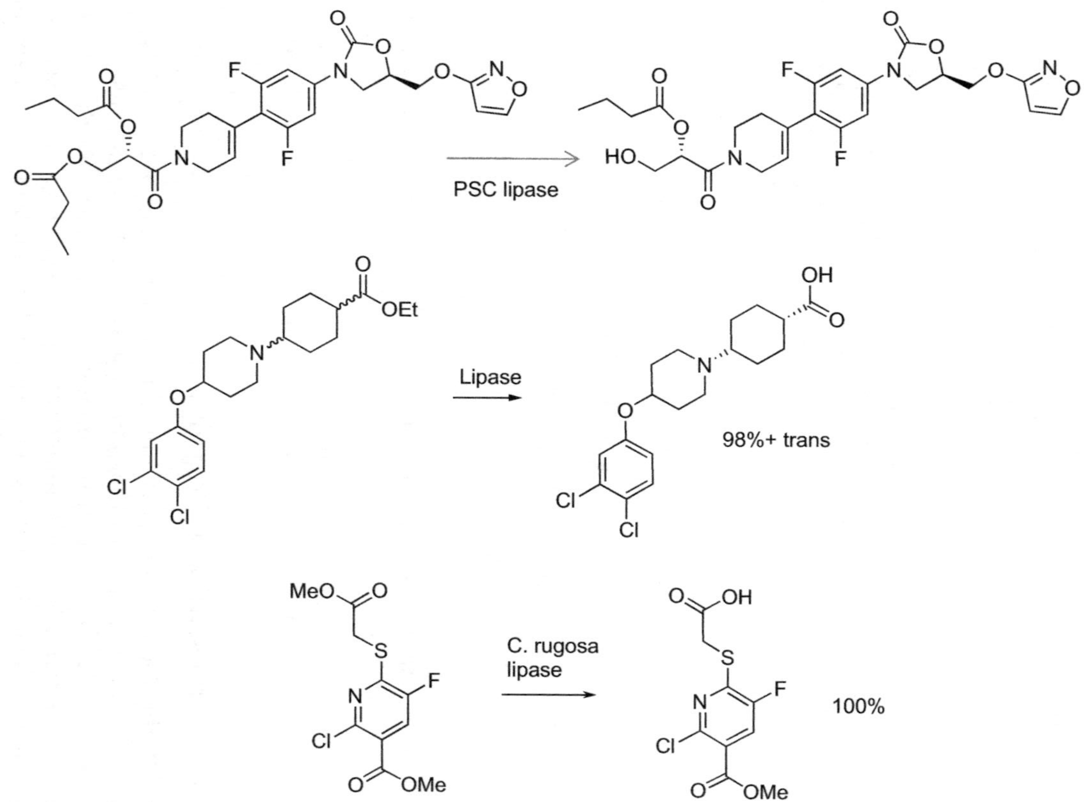

Figure 15.4 Regioselective hydrolysis using hydrolase enzymes.

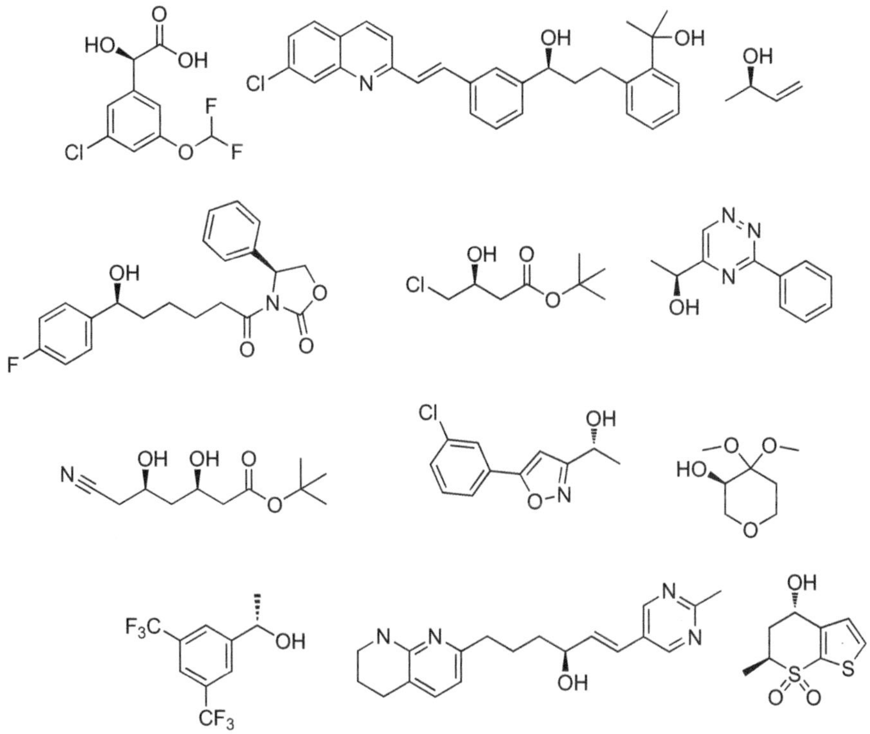

Figure 15.5 A selection of chiral alcohols produced *via* enzyme-catalysed reduction of the corresponding ketones.

class as a synthetic tool has benefited greatly from the ability to mutate enzymes—especially in switching the enantiopreference to give a range of *R* as well as *S* selective enzymes.[16] Figure 15.6 shows a selection of chiral amines synthesized using transaminases.

15.2.4 Cytochrome P450s

Cytochrome P450 (CYP) enzymes are mono-oxidases that catalyse oxidation using O_2 as the terminal oxidant. They are responsible for starting the process of activation, metabolism and excretion of drugs, thus understanding of drug–CYP interactions is fundamental in the pharmaceutical discovery and development process. Traditionally studied with liver microsomes, many human, dog and rodent CYPs have now been sequenced and over-expressed in *E. coli*, yeasts and insect cells for drug metabolism and pharmacokinetics (ADME/DMPK) studies. Many new engineered CYPs are also being developed for use in synthesis.[17] Increasingly, non-mammalian bacterial CYPs are being evolved and developed for synthetic applications in an attempt to overcome the inherent instability and complexity issues found with mammalian CYPs. Two examples are P450cam[18] and P450 BM3

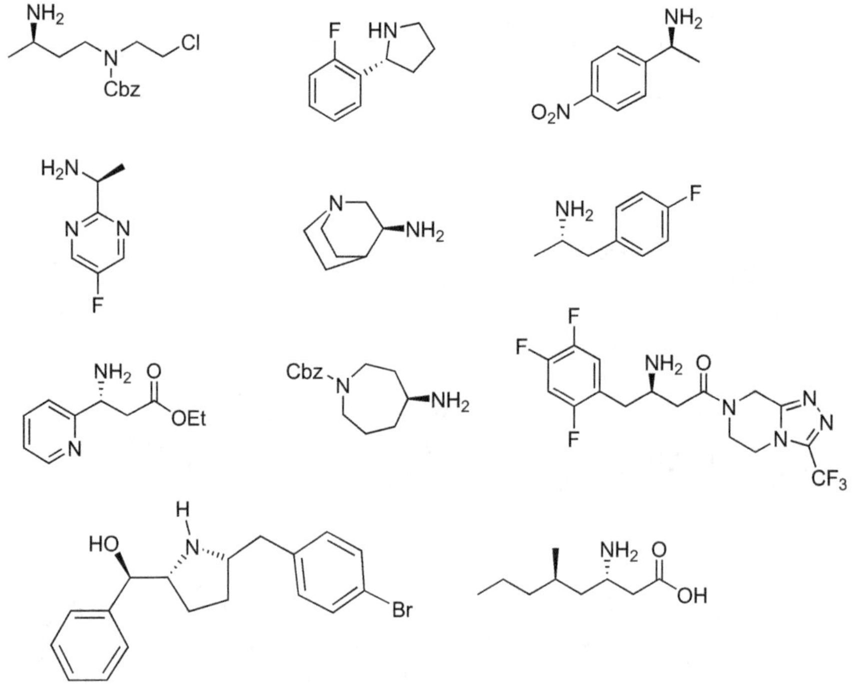

Figure 15.6 Chiral amines accessible from ketones using ω-transaminases and amine donors (typically alanine, isopropyl amine, α-methylbenzylamine).

mutants.[19] Figure 15.7 shows some typical CYP transformations catalysed by P450 cam mutants. As well as use in ADME/DMPK studies, mutant CPYs are increasingly being used to enhance compound collections. These enzymes can produce unusual hydroxylated products, often single enantiomers, that are difficult or impossible to prepare using traditional organic synthesis techniques. These in turn can be converted to novel fluorinated compounds using chemical fluorinating reagents.[20]

15.3 A Glimpse at Synthetic Biology and Pharmaceutical Synthesis

This topic will be only briefly highlighted here as synthetic biology (SB) is the topic of the previous chapter (Chapter 14) in this edition. Amyris and Sanofi have commercialised a semi-synthetic artemisinic acid process to enable the cost-effective production of artemisinin, a key API for antimalarial combination therapies, by inserting the artemisininic acid pathway into *S. cerevisiae*.[21] For discovery research, SB is starting to provide exciting new pathways into unique and novel secondary metabolites as possible new drugs or lead compounds for semi-synthetics. Walker *et al.* have demonstrated the ability of two polyketide synthase systems to accept fluoroacetate in place of acetate leading to the production of a number of novel

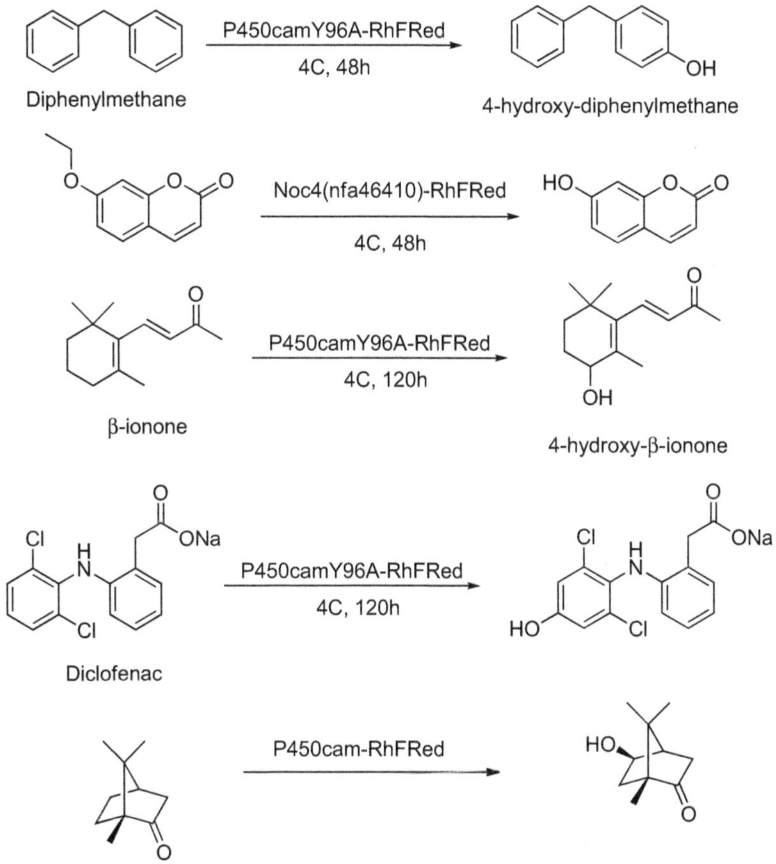

Figure 15.7 Representative transformations undertaken by CYP mutants.[18]

fluorinated natural products.[22] Epigenetic manipulation of genomes is becoming increasingly used to alter secondary metabolite profiles to give rise to novel chemical structures. For example, *Penicillium citreonigrum* chemically challenged with 5-azacytidine (a DNA methyltransferase inhibitor) gave a complex mixture of new metabolites, some of which were active against *Staphylococcus* and *Candida* strains.[23]

15.4 Chapter Summary

- The tools of modern molecular biology have led to an explosion in the number and range of biocatalysts available for use in the pharmaceutical industry.
- Biocatalysis is an underpinning technology to safe and low environmental impact synthesis of APIs and intermediates.
- Many classes of enzyme of interest to the synthetic organic chemist are now available commercially.

- Bespoke mutant enzymes for synthesis can now be rapidly developed.
- Synthetic biology is starting to make an impact in the pharmaceutical manufacturing and discovery arenas.

References

1. M. C. Bryan, B. Dillon, L. G. Hamann, G. J. Hughes, M. E. Kopach, E. A. Peterson, M. Pourashraf, I. Raheem, P. Richardson, D. Richter and H. F. Sneddon, *J. Med. Chem.*, 2013, **56**, 6007.
2. S. D. Roughley and A. M. Jordan, *J. Med. Chem.*, 2011, **54**, 3451.
3. M. T. Reetz, *J. Am. Chem. Soc.*, 2013, **135**, 12480.
4. G. A. Behrens, A. Hummel, S. K. Padhi, S. Schätzle and U. T. Bornscheuer, *Adv. Synth. Catal.*, 2011, **353**, 2191.
5. S. Jennewein, M. Schürmann, M. Wolberg, I. Hilker, R. Luiten, M. Wubbolts and D. Mink, *Biotechnol. J.*, 2006, **1**, 537.
6. C. K. Savile, J. M. Janey, E. C. Mundorff, J. C. Moore, S. Tam, W. R. Jarvis, J. C. Colbeck, A. Krebber, F. J. Fleitz, J. Brands, P. N. Devine, G. W. Huisman and G. J. Hughes, *Science*, 2010, **329**, 305–309.
7. G. W. Huisman and S. J. Collier, *Curr. Opin. Chem. Biol.*, 2013, **17**(2), 284.
8. A. J. Ruff, A. Dennig and U. Schwaneberg, *FEBS J.*, 2013, **280**(13), 2961; M. Goldsmith and D. S. Tawfik, *Curr. Opin. Struct. Biol.* , 2012, **22**(4), 406–412; H. S. Toogood and N. S. Scrutton, *Catal. Sci. Technol.*, 2013, **3**, 2182.
9. N. J. Turner and E. O'Reilly, *Nat. Chem. Biol.*, 2013, **9**, 285.
10. S. Kim, S. Jiménez-González and B. E. Dale, *Int. J. LCA*, 2009, **14**, 392.
11. *Comprehensive Chirality*, ed E. Carreira and H. Yamamoto, Elsevier, 1st edn, 2012.
12. M. Ahmed, T. Kelly and A. Ghanem, *Tetrahedron*, 2012, **68**, 6781; H. J. Lee, K. Han, M.-J. Kim and J. Park, *Eur. J. Org. Chem.*, 2010, **6**, 999.
13. G. W. Huisman, J. Liang and A. Krebber, *Curr. Opin. Chem. Biol.*, 2009, **14**, 1.
14. H. Groeger, W. Hummel, S. Borchert and M. Krausser, *Catalysis in Organic Synthesis*, Wiley, (3rd edn), 2011, p. 1037.
15. W. Kroutil, E.-M. Fischereder, C. S. Fuchs, H. Lechner, F. G. Mutti, D. Pressnitz, A. Rajagopalan, J. H. Sattler, R. C. Simon and E. Siirola, *Org. Process Res. Dev.*, 2013, **17**, 751.
16. M. S. Humble, K. E. Cassimjee, V. Abedi, H.-J. Federsel and P. Berglund, *ChemCatChem*, 2012, **4**, 1167.
17. J. M. CaswelL, M. O'Neill, S. J. Taylor and T. S. Moody, *Curr. Opin. Chem. Biol.*, 2013, **17**, 1.
18. A. Robin, V. Köhler, A. Jones, A. Ali, P. D. Kelly, E. O'Reilly, N. J. Turner and S. L. Flitsch, *Beilstein J. Org. Chem.*, 2011, **7**, 1494.
19. A. M. Sawayama, M. M. Chen, P. Kulanthaivel, M.-S. Kuo, H. Hemmerle and F. H. Arnold, *Chem. – Eur. J.*, 2009, **15**, 11723.
20. A. Rentmeister, F. H. Arnold and R. Fasan, *Nat. Chem. Biol.*, 2009, **5**, 26.
21. C. J. Paddon, P. J. Westfall, D. J. Pitera, K. Benjamin, K. Fisher, D. McPhee, M. D. Leavell, A. Tai, A. Main, D. Eng, D. R. Polichuk,

K. H. Teoh, D. W. Reed, T. Treynor, J. Lenihan, M. Fleck, S. Bajad, G. Dang, D. Dengrove, D. Diola, G. Dorin, K. W. Ellens, S. Fickes, J. Galazzo, S. P. Gaucher, T. Geistlinger, R. Henry, M. Hepp, T. Horning, T. Iqbal, H. Jiang, L. Kizer, B. Lieu, D. Melis, N. Moss, R. Regentin, S. Secrest, H. Tsuruta, R. Vazquez, L. F. Westblade, L. Xu, M. Yu, Y. Zhang, L. Zhao, J. Lievense, P. S. Covello, J. D. Keasling, K. K. Reiling, N. S. Renninger and J. D. Newman, *Nature*, 2013, **496**, 528, Apr 25.

22. C. M. Walker, B. M. Thuronyi, L. K. Charkoudian, B. Lowry, C. Khoslaand and M. C. Chang, *Science*, 2013, **341**, 1089.
23. X. Wang, J. G. Sena Filho, A. R. Hoover, J. B. King, T. K. Ellis, D. R. Powell and R. H. Cichewicz, *J. Nat. Prod.*, 2010, **73**, 942.

CHAPTER 16

Base Metals in Catalysis: From Zero to Hero

J. MAES, E. A. MITCHELL AND B. U. W. MAES*

University of Antwerp, Department of Chemistry, Groenenborgerlaan 171, 2020 Antwerpen, Belgium
*Email: bert.maes@uantwerpen.be

16.1 Base Metals: "How Can We Serve You"?

The use of catalysis as a tool for organic synthesis is one of the twelve guiding principles that lie at the foundation of green chemistry.[1] Currently, more than 85% of all chemical products worldwide, both bulk and fine chemicals, are made with the aid of a metal catalyst.[2] Until now, especially in the pharmaceutical industry, most of these processes use precious metals to perform chemical transformations that are normally impossible to realize in their absence. The limited abundance and reserves of these noble metals can therefore have a significant impact on the efficiency of organic syntheses and the cost of the catalysts and the chemicals produced *via* catalysis (see Table 16.1). Even for those elements where reserves are predicted to last for another 100 years, stability in the price of these elements is by no means certain based on current rates of extraction. In addition, geopolitical uncertainty in areas currently mining a major deposit of a specific precious metal, as well as potential market manipulation to limit the amount exported, can easily cause global shortages on the international market. This is often considered to represent a more acute threat to the chemical industry than world reserve depletion and is thus cause for major concern.[3] This is

RSC Green Chemistry No. 46
Green and Sustainable Medicinal Chemistry: Methods, Tools and Strategies for the 21st Century Pharmaceutical Industry
Edited by Louise Summerton, Helen F. Sneddon, Leonie C. Jones and James H. Clark

Published by the Royal Society of Chemistry, www.rsc.org

Table 16.1 Number of years remaining of (precious) metal reserves if consumption continues at present rate.[6]

Metal	Remaining years until depletion of known reserves[a]
Aluminium	100–500
Chromium	100–500
Iron	>500
Cobalt	50–100
Nickel	50–100
Zinc	50–100
Ruthenium	5–50
Rhodium	5–50
Palladium	50–100
Silver	5–50
Platina	5–50
Gold	5–50

[a]Based on current rate of extraction.

particularly true when one considers the synthesis of the complex chemical entities commonly produced by the pharmaceutical industry. In this instance, it is not straightforward to identify an alternative to a precious metal catalyst for a particular synthesis, nor is it evident which of these alternatives are safe, provide the appropriate chemical selectivity and do not require partial or full redrawing of the chemical production route.[4] In addition, even small changes in the manufacturing route of an API require the reopening of the associated registration files, causing additional costs.[5]

Classical metal reclaiming processes are currently applied in the pharmaceutical industry, but these methods are not necessarily performed with sustainability in mind. Obviously, no recovery is worse. Furthermore, the ligands of a given catalyst, which are often more expensive per mol than the metal itself, are lost during this process. As a result, significant research efforts have been made in the last decade to develop sustainable separation and recovery techniques for precious metals.[6] Another strategy to avoid the disadvantages of uncertain precious metal reserves is to shift away from their use towards more readily available and cheaper base metals in catalysis. This often serves major advantages beyond reduced cost and availability: many base metals exhibit low toxicity and are environmentally benign. However, it is important to note that toxicity is related to a metal species as a whole, including the ligands in the metal's coordination sphere, and one should therefore be careful in generalizing the toxicities of specific metals. An excellent example of this concept is iron. While iron is generally considered to be a non-toxic transition metal, $Fe(CO)_5$ possesses a high acute toxicity.[7]

Base metals are currently invaluable to the global economy because of their ubiquity and utility. However, it is not always clear which metals are actually considered to be base metals. The narrowest definition covers only those metals that react with aqueous HCl, including iron, nickel, lead, cobalt and zinc. A broader definition of base metals includes those metals that are in general easily oxidized. As one typically wants to refer to inexpensive and more abundant metals when using the term 'base metals', the latter definition is preferred. This definition is obviously not limited to transition metals. The transition metals that are considered to belong to the category of 'base metals' are therefore those with non-noble behavior.

It is in fact these metals that are generally seen as the most promising substitutes for precious metals in the area of C–C and C–X (heteroatom) bond formation *via* transition-metal catalysis. In this chapter, we focus on C–N bond forming reactions to exemplify the importance and future potential of base metal catalysis for API synthesis. The choice is based on the primary role of nitrogen-containing and amino-substituted heterocycles in the fine chemicals industry. The synthesis of C–N bonds using transition metal catalysis starting from activated substrates [C–LG bond (LG = leaving group)] has matured significantly in the past 20 years (Scheme 16.1, entry 1) by the development of the Buchwald–Hartwig reaction (Pd catalysis) and major improvements in Ullmann type reactions (Cu-catalysis). Even base metals (mainly copper) can now efficiently be used in these transformations.[8] C–N bond formation *via* direct functionalization of a C–H bond represents the next step in the evolution of this type of transformation and has proven to be a very challenging, "hot" topic in contemporary homogenous catalysis research. Although the field is still in its infancy, good methods using base metals have already been developed. Direct C–H bond functionalization (direct amination) possesses major advantages over the above-mentioned classical methods since no preactivation (introducing an LG in the substrate) is needed, thus reducing the production of waste (Scheme 16.1, entry 2). However, C–H bonds are stronger than C–LG bonds and are ubiquitous in organic molecules, creating serious new challenges in terms of activity and selectivity compared to those catalytic processes starting from C–LG bonds. Furthermore, a stoichiometric oxidant is required for this type of transition metal-catalyzed reaction in order to make the process catalytic. Common examples of oxidants are oxygen, *tert*-butyl hydroperoxide (TBHP), oxone, phenyliodonium diacetate (PIDA), benzoquinone (BQ) and 2,3-dichloro-5,6-dicyano-1,4-benzoquinone (DDQ). It is clear that the chosen oxidant for a given direct C–N bond formation is crucial since it will

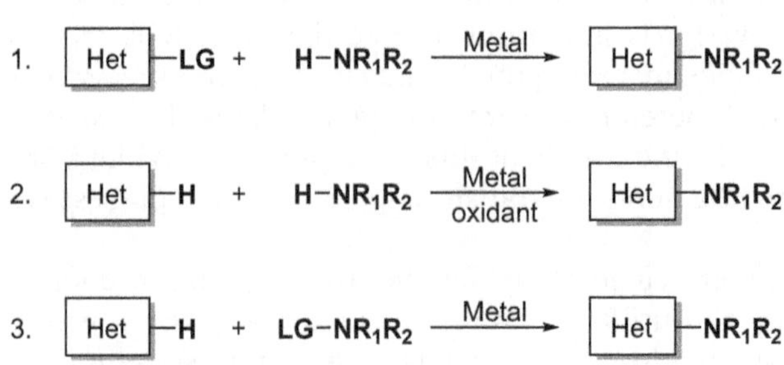

Scheme 16.1 C–N bond formation strategies.

determine whether the approach is beneficial, both in terms of economy and sustainability, compared to the classical transition metal-catalyzed approach (using C–LG). It is also possible for the leaving group to be a component of the nucleophile involved in C–N bond formation, but procedures involving nucleophile activation are not expected to perform better in terms of sustainability than methods that use substrate activation (Scheme 16.1, entry 3). Such an approach might however be preferred for purely synthetic reasons, such as the availability of the activated substrate. In this work, we will highlight the use of molecular oxygen as an oxidant and instances where both the substrate and nucleophile are not preactivated, often labeled as cross-dehydrogenative coupling reactions. Oxygen is the most sustainable oxidant on earth, as it is inexpensive and possesses an environmentally benign character. Moreover, it is abundantly available and water is generated as the sole by-product of its use.[9] As the oxidant in C–N bond formation *via* C–H activation, oxygen participates in an oxidase type mechanism, in which it functions to regenerate the base metal catalyst by returning it to its initial oxidation state (Scheme 16.2).[10] Nevertheless, it is important to state that the use of oxygen in an industrial process, rather than on a lab scale, is not without challenges as it can cause fires and explosions. A mixture of oxygen and nitrogen is thus often used in large-scale batch production to obtain good control over the experiment and to ensure safety. Moreover, the rather low solubility of oxygen in organic solvents creates challenges for reactions with a rate-limiting oxidation step where the concentration of oxygen has a direct effect on the efficiency of the catalysis.[11] The use of flow reactors to produce chemicals in high volume *via* continuous processes has brought about new opportunities to tackle these issues. Flow chemistry makes increased mixing of gas and liquid phases possible, which serves to improve reaction rates, and the limited reaction volumes involved in this technology inherently allow one to work with dangerous gases in a safer way.[12]

Throughout this chapter, we will focus on base metal-catalyzed reactions that create C–N bonds *via* C–H activation using amines as reagents and oxygen as the oxidant. This work does not aim to be comprehensive, but merely highlights some selected examples that illustrate the current 'state-of-the-art' in this field.

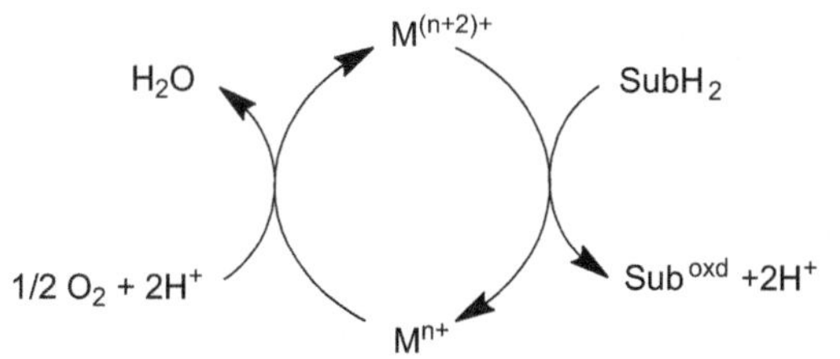

Scheme 16.2 General oxidase-type mechanism.

16.2 Base Metal Catalysis as a Sustainable Toolbox in Modern Chemistry: The Direct Amination Case

One of the first examples of a direct amination reaction involving base metal catalysis was published by Buchwald's group in 2008.[13] 2-Arylbenzimidazoles could be made *via* a C–N cyclization reaction of *N*-phenylarylimidamide with $Cu(OAc)_2$ as the catalyst and acetic acid as the additive of choice (Scheme 16.3, (**1**)). Several functional groups were well tolerated by the protocol; moreover the method could be extended to *N*-methylated 2-arylbenzimidazole and 2-alkylbenzimidazole synthesis in good yields.

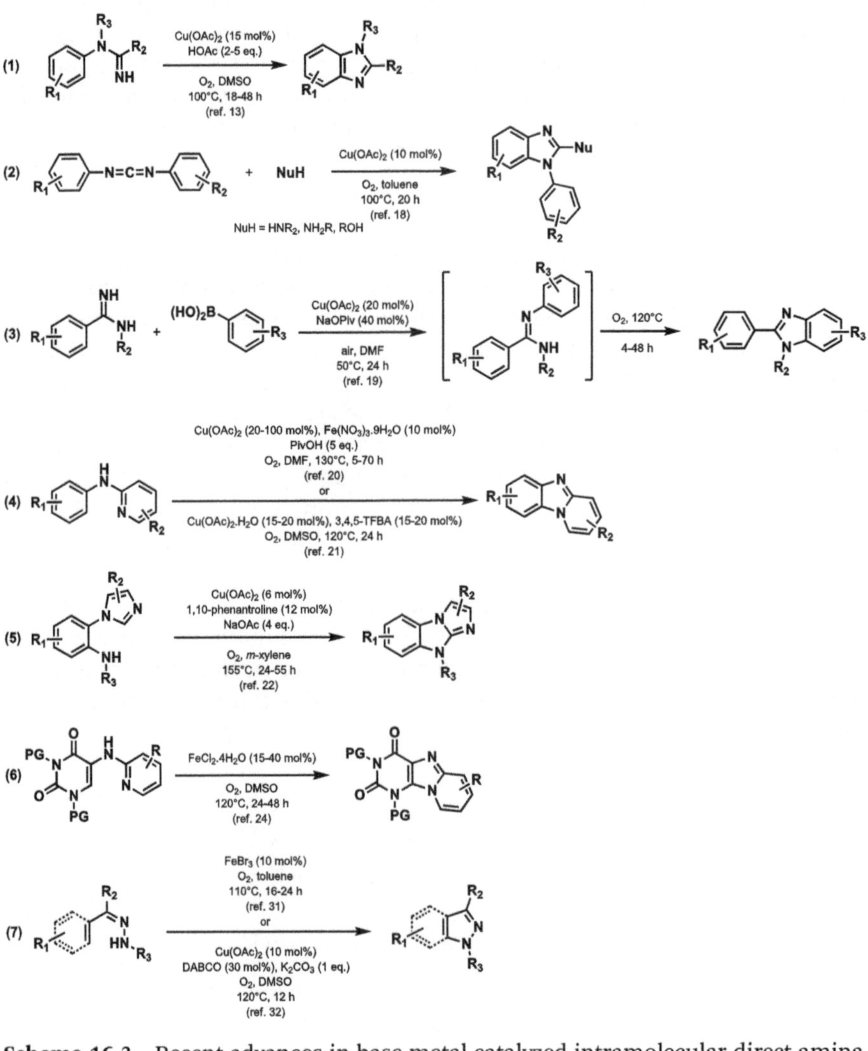

Scheme 16.3 Recent advances in base metal catalyzed intramolecular direct amination reactions.

For the latter type, it was only a *tert*-butyl group that was tolerated, as other alkyl groups in the substrate gave complete decomposition. The mechanism of this reaction was recently investigated by Fu[14] and Li.[15] Li *et al.* concluded that the Cu(II)-assisted concerted metalation–deprotonation (CMD) pathway is the most favorable pathway for this reaction.[14] The mechanism consists of five steps: (1) coordination, (2) concerted metalation–deprotonation with Cu(II), (3) oxidation of the Cu(II) intermediate to Cu(III) followed by (4) deprotonation and finally (5) reductive elimination of the Cu(III) species. An OAc-assisted six membered ring transition state was suggested in accordance with what was introduced by Echavarren[16] and Fagnou[17] for Pd-catalyzed C–H activation pathways. The C–N cyclization was later used by the groups of Bao[18] and Zhu[19] in a tandem and catalytic one-pot reaction, respectively, for the synthesis of C-2 functionalized benzimidazoles. Bao *et al.* designed a nucleophilic addition reaction on (a)symmetrically substituted diarylcarbodimiides followed by a copper catalyzed intramolecular direct C–N bond formation reaction (Scheme 16.3, (**2**)). In general, high yields were obtained with electron-donating substituents in the 4-position of the diarylcarbodiimide and comparatively lower yields were obtained for substituents in the 2- and 3-positions. Zhu *et al.* created an intermediate for direct amination upon reaction of a substituted arylamidine with an arylboronic acid bearing at least one hydrogen in the *ortho*-position (Scheme 16.3, (**3**)). This Chan–Lam–Evans *N*-arylation could be executed under catalytic copper conditions under air where, after changing the atmosphere for pure oxygen and elevating the temperature up to 120 °C, the corresponding substituted benzimidazoles could be obtained. An important extension of this type of chemistry is the synthesis of C(2)-N annulated benzimidazoles. Simultaneously, the groups of Zhu[20] and Maes[21] described a copper-based protocol for the synthesis of pyrido[1,2-*a*]benzimidazoles (Scheme 16.3, (**4**)). In both cases, copper acetate was chosen as the catalyst, but in the first case an iron co-catalyst and stoichiometric acid additive was needed to obtain good conversions. In the latter case, only a catalytic amount of acid (3,4,5-trifluorobenzoic acid) was proven to be sufficient to bring the reaction to completion and good to excellent yields were obtained. Fu *et al.* demonstrated that base metal catalysis could also be used for the synthesis of imidazobenzimidazole derivatives (Scheme 16.3, (**5**)).[22] A combination of $Cu(OAc)_2$ and 1,10-phenantroline was chosen as the catalyst system and NaOAc was added as a base to facilitate the proton abstraction in the first step of the mechanism. With slightly altered reaction conditions, Kaliappan's group could make similar reaction products. However, in this case, an auto-tandem C–X/C–H amination allowed the additional introduction of an *N*-aryl ring.[23] Remarkably, an excess of oxygen completely shuts down the reaction and full conversion can only be obtained when the reaction is executed in a closed vessel. This can be explained by the different oxidation state of the transition metal in each catalytic cycle. The active copper species for the C–X amination is Cu(I) whereas Cu(II) is necessary for the direct C–H amination. DMF as solvent reduces the precatalyst to Cu(I) for the C–X amination. On the other hand, the limited amount of oxygen present in the closed vial oxidizes Cu(I) to Cu(II) for the direct C–H amination. Recently, Maes' group reported the first example of an iron-catalyzed direct amination reaction with a non-preactivated *N*-nucleophile (Scheme 16.3, (**6**)).[24] Up to now, known Fe-catalyzed sp^3 and sp^2 C–H amination reactions generally involve nitrenoid formation. These intermediates are generated from preactivated primary amines (*e.g.* from azides,[25] PhI=NTs,[26] PhNO and PhNHOH,[27] by *in situ* oxidation of a primary amine with *N*-bromosuccinimide (NBS)[28] or $PhI(OAc)_2$,[29] or by ring opening of an 2*H*-azirine[30]). However, the stoichiometric preactivation of the nitrogen reagent is still a drawback from the perspective of sustainability. The new protocol was used for the synthesis of substituted pyrido[1,2-*e*]purines, which can potentially interact selectively with enzymes and receptors. In contrast to the previous examples of direct C–H amination in this chapter, the proposed mechanism proceeds *via* a radical rather than an ionic pathway, as illustrated by experiments in the presence of radical inhibitors. The method possesses excellent functional group compatibility and the chemoselectivity towards halogens allows for further functionalization of the annulated ring. The mechanism is highly dependent on the catalyst selected, as demonstrated by the work of Bao[31] and Jiang.[32] They both reported the synthesis of indazoles *via* an intramolecular C–H amination reaction under iron and copper catalysis, respectively. When the reaction is iron catalyzed, the mechanism is claimed to proceed *via* a radical pathway whereas the copper catalyzed reaction is considered to follow an ionic pathway, as illustrated by experiments with radical scavengers.

Besides the intramolecular direct C–N bond formation, base metals can also be used for C–N bond formation in an intermolecular manner. Azoles such as benzimidazoles, (benz)oxazoles, and (benzo)thiazoles are important heterocyclic cores and can be found in a variety of different biologically active and natural compounds. To tune the properties of these scaffolds, functionalization of the skeleton is essential. Since 2009, several research groups have investigated the direct functionalization of the C-2 position of azoles with amines *via* base metal catalysis.[33–38] Mori *et al.* described first the intermolecular C–H amination of azoles with oxygen as the terminal oxidant (Scheme 16.5, (**1**)).[33] The catalyst system $Cu(OAc)_2/PPh_3$ was found to be optimal and could be applied to benzo-thiazoles, -oxazoles and -imidazoles. The proposed mechanism is shown in Scheme 16.4 and starts with replacing the sp^2 C–H bond with Cu(II) (no details on the way the C–H bond is activated are given). Under basic conditions a ligand exchange takes place, which can upon reductive elimination give the coupled product. The obtained copper species is re-oxidized by oxygen to complete the catalytic cycle.

Similar transformations are described by Schreiber[34] and Duan[35] (see Scheme 16.5, (**2–4**)) where *N*-nucleophiles are coupled under oxidative conditions *via* copper catalysis. The protocols are applicable to not only secondary amines but also cyclic secondary amides, primary amides and sulfonamides or *via* decarboxylative coupling with formamides under acidic

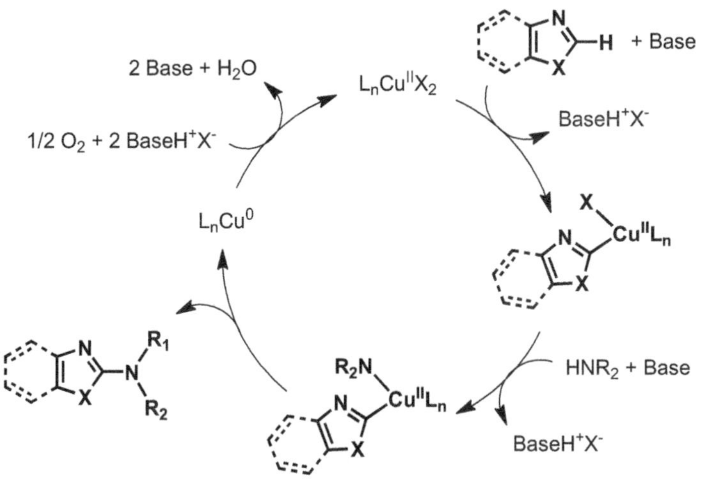

Scheme 16.4 Proposed catalytic cycle for the direct C(2)-amination of azoles.

conditions. Miura and Bolm developed conditions where sulfoximines were used as the *N*-nucleophile in the *N*-arylation (Scheme 16.5, (**5**)).[36] Mechanistic experiments for the sulfoximination revealed that in this specific case oxygen has a dual role (not only as terminal oxidant) since only traces of end compound were obtained when the reaction was executed with stoichiometric copper under inert atmosphere. Huang *et al.* developed an oxidative amination reaction of azoles which can even use tertiary amines as nitrogen source (Scheme 16.5, (**6**)).[37] The key step in the reaction is the copper catalyzed formation of an iminium intermediate, which upon hydrolysis gives the corresponding copper amide. This copper amide coordinates with the azole nitrogen followed by deprotonation/rearrangement and subsequently reductive elimination yielding the reaction product and reduced copper. The same research group further investigated this type of chemistry and developed a much milder protocol based on cooperative catalysis where organocatalysis is combined with copper catalysis for the aerobic oxidative C–H amination of benzoxazoles.[38] Aldehyde catalysis was used to form an aminal *in situ* from a secondary amine, which upon hydrolysis gave a similar copper–amide species as used in their initial publication.

16.3 Chapter Summary

- Limited abundance and reserves of noble metals on earth can have a significant impact on the efficiency of organic syntheses.
- Base metals serve major advantages, such as reduced cost, high abundance, low toxicity and their environmentally benign character.
- Base metals are generally seen as future substitutes for precious metals in the area of C–C and C–X (heteroatom) bond formation *via* transition-metal catalysis.

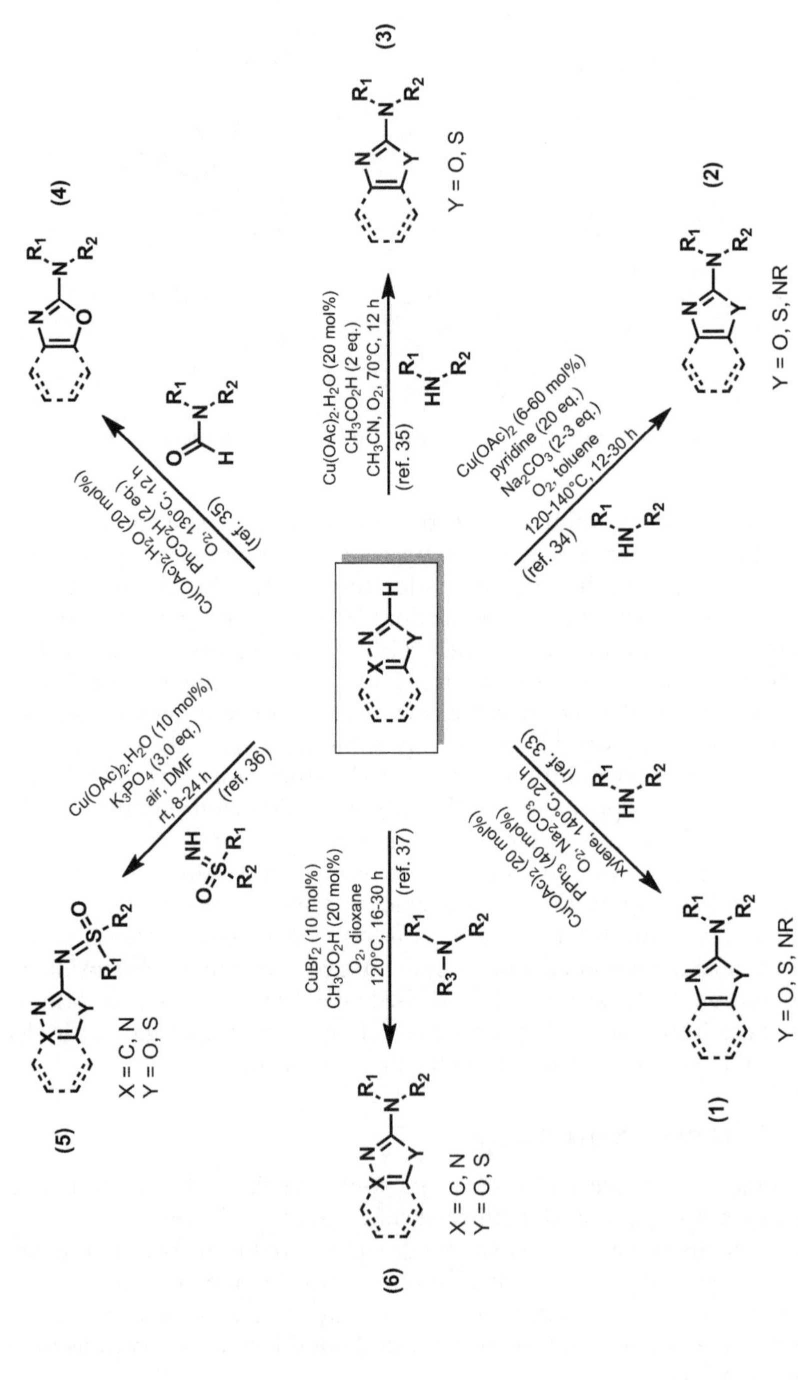

Scheme 16.5 Recent advances in the base metal catalyzed intermolecular direct amination of azoles.

- Direct C–H amination requires no preactivation and therefore generates less waste.
- Molecular oxygen is the oxidant of choice due to its environmentally benign character.

References

1. (a) P. T. Anastas, J. C. Warner in *Green Chemistry: Theory and Practice*, Oxford University Press, New York (USA), 1998; (b) J. Clark, D. Macquarrie, M. Gronnow and V. Budarin in *Process Intensification for Green Chemistry: Engineering Solutions for Sustainable Chemical Processing*, ed K. Boodhoo and A. Harvey, John Wiley & Sons, Ltd, Chichester (UK), 2013, p. 33.
2. J. Heveling, *J. Chem. Educ.*, 2012, **89**, 1530.
3. European Commission Enterprise and Industry, Critical Raw Materials for the EU, Report of the Ad-hoc Working Group on Defining Critical Raw Materials, 2010, http://ec.europa.eu/enterprise/policies/raw-materials/files/docs/report-b_en.pdf.
4. M. Butters, D. Catterick, A. Craig, A. Curzons, D. Dale, A. Gillmore, S. P. Green, I. Marziano, J.-P. Sherlock and W. White, *Chem. Rev.*, 2006, **106**, 3002.
5. J. Curran in *Active Pharmaceutical Ingredients: Development, Manufacturing, and Regulation*, ed S. H. Nusim, Taylor & Francis Group, LLC, Florida (USA), 2nd edn, 2005, p. 167.
6. A. J. Hunt, T. J. Farmer and J. H. Clark in *Elemental Recovery and Sustainability*, ed A. J. Hunt, The Royal Society of Chemistry, Cambridge (UK), 2013, p. 1.
7. Iron(0) Pentacarbonyl (2013). In Kaleidos [Dbase]. Geel, Belgium: BIG vzw, online: http://www.big.be.
8. (a) A. R. Muci and S. L. Buchwald in *Palladium Catalysts for C-N and C-O Bond Formation in Cross-Coupling Reactions: A Practical Guide*, ed N. Miyaura, Springer-Verlag Berlin, Heidelberg (Germany), 2002, p. 133; (b) P. Das, D. Sharma, M. Kumar and B. Singh, *Curr. Org. Chem.*, 2010, **14**, 754.
9. Z. Shi, C. Zhang, C. Tang and N. Jiao, *Chem. Soc. Rev.*, 2012, **41**, 3381.
10. S. S. Stahl, *Angew. Chem., Int. Ed.*, 2004, **43**, 3400.
11. R. Battino, T. R. Rettich and T. Tominaga, *J. Phys. Chem. Ref. Data*, 1983, **12**, 163.
12. (a) S. G. Newman and K. F. Jensen, *Green Chem.*, 2013, **15**, 1456; (b) G. Sipos, V. Gyollai, T. Sipőcz, G. Dormán, L. Kocsis, R. V. Jones and F. Darvas, *J. Flow Chem.*, 2013, **3**, 51.
13. G. Brasche and S. L. Buchwald, *Angew. Chem., Int. Ed.*, 2008, **47**, 1932.
14. S.-Y. Tang, T.-J. Gong and Y. Fu, *Science China-Chemistry*, 2013, **56**, 619.
15. J. Li, H. Gu, C. Wu and L. Du, *Dalton Trans.*, 2014, **43**, 16769.
16. D. Garcia-Cuadrado, F. Maseras and A. M. Echavarren, *J. Am. Chem. Soc.*, 2006, **128**, 1066.

17. D. Lapointe and K. Fagnou, *Chem. Lett.*, 2010, **39**, 1118.
18. H.-F. He, Z.-J. Wang and W. Bao, *Adv. Synth. Catal.*, 2010, **352**, 2905.
19. J. Li, S. Bénard, L. Neuville and J. Zhu, *Org. Lett.*, 2012, **14**, 5980.
20. H. Wang, Y. Wang, C. Peng and Q. Zhu, *J. Am. Chem. Soc.*, 2010, **132**, 13217.
21. K.-S. Masters, T. R. M. Rauws, A. K. Yadav, W. A. Herrebout, B. Van der Veken and B. U. W. Maes, *Chem. – Eur. J.*, 2011, **17**, 6315.
22. X. Wang, Y. Jin, Y. Zhao, L. Zhu and H. Fu, *Org. Lett.*, 2012, **14**, 452.
23. P. Subramanian and K.-P. Kaliappan, *Eur. J. Org. Chem.*, 2014, **27**, 5986.
24. J. Maes, T. R. M. Rauws and B. U. W. Maes, *Chem. – Eur. J.*, 2013, **19**, 9137.
25. (a) Y. Liu and C.-M. Che, *Chem. – Eur. J.*, 2010, **16**, 10494; (b) Y. Liu, J. Wei and C.-M. Che, *Chem. Commun.*, 2010, **46**, 6926; (c) B. J. Stokes, C. V. Vogel, L. K. Urnezis, M. Pan and T. G. Driver, *Org. Lett.*, 2010, **12**, 2884; (d) J. Bonnamour and C. Bolm, *Org. Lett.*, 2011, **13**, 2012; (e) E. R. King, E. T. Hennessy and T. A. Betley, *J. Am. Chem. Soc.*, 2011, **133**, 4917; (f) L. Liang, H. Lv, Y. Yu, P. Wang and J.-L. Zhang, *Dalton Trans.*, 2012, **41**, 1457; (g) Q. Nguyen, T. Nguyen and T. G. Driver, *J. Am. Chem. Soc.*, 2013, **135**, 620.
26. (a) R. Breslow and S. H. Gellman, *J. Am. Chem. Soc.*, 1983, **105**, 6728; (b) S.-Y. Yan, Y. Wang, Y.-J. Shu, H.-H. Liu and X.-G. Zhou, *J. Mol. Catal. A: Chem.*, 2006, **248**, 148; (c) T. M. U. Ton, C. Tejo, S. Tania, J. W. W. Chang and P. W. H. Chan, *J. Org. Chem.*, 2011, **76**, 4894; (d) G.-Q. Chen, Z.-J. Xu, Y. Liu, C.-Y. Zhou and C.-M. Che, *Synlett*, 2011, **8**, 1174–1178.
27. (a) R. S. Srivastava, M. A. Khan and K. M. Nicholas, *J. Am. Chem. Soc.*, 1996, **118**, 3311; (b) R. S. Srivastava and K. M. Nicholas, *J. Am. Chem. Soc.*, 1997, **119**, 3302; (c) R. S. Srivastava, M. A. Khan and K. M. Nicholas, *J. Am. Chem. Soc.*, 2005, **127**, 7278; (d) B. Kalita and K. M. Nicholas, *Tetrahedron Lett.*, 2005, **46**, 1451; (e) R. S. Srivastava, N. R. Tarver and K. M. Nicholas, *J. Am. Chem. Soc.*, 2007, **129**, 15250.
28. Z. Wang, Y. Zhang, H. Fu, Y. Jiang and Y. Zhao, *Org. Lett.*, 2008, **10**, 1863.
29. S. M. Paradine and M. C. White, *J. Am. Chem. Soc.*, 2012, **134**, 2036.
30. S. Jana, M. D. Clements, B. K. Sharp and N. Zheng, *Org. Lett.*, 2010, **12**, 3736.
31. T. Zhang and W. Bao, *J. Org. Chem.*, 2013, **78**, 1317.
32. X. Li, L. He, H. Chen, W. Wu and H. Jiang, *J. Org. Chem.*, 2013, **78**, 3636.
33. D. Monguchi, T. Fuijwara, H. Furukawa and A. Mori, *Org. Lett.*, 2009, **11**, 1607.
34. Q. Wang and S. L. Schreiber, *Org. Lett.*, 2009, **11**, 5178.
35. Y. Li, Y. Xie, R. Zhang, K. Jin, X. Wang and C. Duan, *J. Org. Chem.*, 2011, **76**, 5444.
36. M. Miyasaka, K. Hirano, T. Satoh, R. Kowalczyk, C. Bolm and M. Miura, *Org. Lett.*, 2011, **13**, 359.
37. S. Guo, B. Qian, Y. Xie, C. Xia and H. Huang, *Org. Lett.*, 2011, **13**, 522.
38. Y. Xie, B. Qian, P. Xie and H. Huang, *Adv. Synth. Catal.*, 2013, **355**, 1315.

CHAPTER 17

'Green' and Sustainable Halogenation Processes

GRAHAM PATTISON

Department of Chemistry, University of Warwick, Gibbet Hill Road, Coventry CV4 7AL, UK
Email: graham.pattison@warwick.ac.uk

17.1 Introduction

Halogen atoms have a privileged position in the design of organic molecules. Fluorine atoms appear in a widespread range of pharmaceuticals, agrochemicals and functional materials. The properties of fluorine, being the most electronegative atom, allow it to strongly modulate properties such as pK_a, lipophilicity and polarity, and give improved biological activity and performance.[1] It has been estimated that fluorine is found in at least 25% of all pharmaceuticals, including the world's best-selling drug, Lipitor.[2] Meanwhile, the other halogens provide valuable 'functional handles' when incorporated into organic molecules; their polar bonds to carbon enable chemistry such as nucleophilic displacements and cross-coupling reactions, which allow more complex molecules to be easily built up.

Chemists therefore require a range of reactions to incorporate halogen atoms into diverse organic molecules, and, as so many halogenated products are produced on a large scale, these halogenation reactions need to be carried out in a green and sustainable fashion. However, the halogens often have a reputation for being anything but green. Fluorine has a fearsome reputation as the most reactive element; hydrogen fluoride is an extremely

RSC Green Chemistry No. 46
Green and Sustainable Medicinal Chemistry: Methods, Tools and Strategies for the 21st Century Pharmaceutical Industry
Edited by Louise Summerton, Helen F. Sneddon, Leonie C. Jones and James H. Clark

Published by the Royal Society of Chemistry, www.rsc.org

dangerous and corrosive acid that damages tissues and whose burns can prove fatal, and chlorine gas is highly toxic and was used as a chemical weapon in World War I. Clearly, chemists must be able to handle these materials responsibly, and where possible develop safer alternatives.

This chapter will demonstrate that the reputation of the halogens is sometimes undeserved and that we are now developing modern alternatives to these reagents that allow us to carry out halogenation safely and with minimal impact on the environment. For example, halide salts are readily available, easily handled and safe sources of halogen atoms. Chemists are also establishing techniques to handle the reactive halogen elements safely and cleanly. All these developments will be highlighted in the upcoming pages. A particular emphasis will be placed on fluorine chemistry due to its ubiquitous position in useful molecules, and the specific challenges its introduction poses. This represents a personal selection of highlights in this field, and many other important areas could not be covered in a brief review like this. Hopefully this should provide a flavour of the possibilities of this rapidly advancing research area and stimulate new advances in the field. A summary table is included (Table 17.1) containing the key advantages and disadvantages of various fluorinating agents.

17.2 Electrophilic Fluorination

The majority of fluorination processes developed to date are electrophilic in nature and a range of organic functional groups can be fluorinated in this way, including aromatic rings, alkenes and enol derivatives. The ideal reagent to achieve this chemistry in terms of atom-economy and waste products generated would be fluorine gas, however its extreme reactivity and difficulties in its handling make its use too challenging for the majority of chemists. A range of alternative electrophilic fluorination reagents, often containing N–F bonds, including Selectfluor, *N*-fluorobenzenesulfonimide, and *N*-fluoropyridinium salts, have been developed that are more easily handled than fluorine. These N–F reagents see widespread utility, for example in medicinal chemistry, however they are not ideal for use in large-scale processes. They are expensive, produce stoichiometric quantities of amine by-products and are ultimately derived from elemental fluorine by fluorination of the corresponding amine.

A strategy to safely utilise fluorine gas is therefore needed if we are to carry out electrophilic fluorination on a large scale. When handled as a 10% dilution of fluorine in nitrogen gas, the reactivity of fluorine is significantly perturbed to the level where it can be safely stored and handled, yet still participates in important fluorination reactions. To further improve the handling of fluorine, Chambers and Sandford in Durham have been developing microreactor systems which allow fluorination of a range of systems in flow.[3] This strategy minimises the concentration of fluorine present at any one time, and by passing the product stream though a series of water and sodalime scrubbers, allows any unreacted fluorine or hydrogen fluoride

Table 17.1 Key advantages and disadvantages of various fluorinating agents.

Fluorinating agent	What is it used to fluorinate?	Why is it green? Advantages	Why is it not green? Disadvantages
Elemental fluorine	Carbonyl compounds Aromatics C–H bonds	Excellent atom economy	Highly reactive, toxic Specialist handling needed Often generates HF by-products
HF	Halide Displacements (S_N2)	Excellent atom economy	Extremely corrosive and hazardous
Fluoride salts (*e.g.* KF, CsF)	Halide Displacements (S_N2)	Safe and easy to handle Excellent atom economy Few by-products produced	Very hygroscopic but often needs to be dry for good reactivity
N-F reagents (*e.g.* Selectfluor, NFSI)	Carbonyl Compounds Aromatics Alkenes Thioethers Organometallic reagents (*e.g.* organotin)	Safe and easy to handle Low toxicity	Derived from elemental fluorine Expensive Amine waste generated
Ruppert reagent (CF_3SiMe_3)	Carbonyl compounds (nucleophilic) Aromatics	Safe and easy to handle	Silicon-containing waste generated Expensive
Electrophilic trifluoromethylation reagents (*e.g.* Togni)	Carbonyl compounds (electrophilic) Aromatics Alkenes Alcohols/sulfides	Few other ways of doing many of these transformations	Iodoarene waste generated Expensive Reports of explosions
DAST/SF_4	Alcohol $\rightarrow$ CF Carbonyl $\rightarrow$ CF_2	Few other ways of doing many of these transformations	Corrosive Explosive byproducts at high temperatures
Fluoroform (CHF_3)	Carbonyl compounds	Using byproduct of Teflon production, which otherwise needs to be disposed of	Very potent greenhouse gas Normally incinerated producing H_2O, CO_2, HF
Trifluoroacetate salts	Aromatics Carbonyl compounds	CO_2 is only byproduct produced on decarboxylation	CO_2 is greenhouse gas Produced using corrosive HF

by-products to be safely removed and neutralised. The use of flow reactors also allows further reaction steps to be introduced in an in-line and continuous fashion. To demonstrate this, a range of 1,3-diketones was first fluorinated in a nickel–hastelloy microreactor, then hydrazine was introduced through a T-piece, which gave the corresponding fluorinated pyrazole derivative **1** in high yield (Scheme 17.1).

N–F reagents such as Selectfluor have increased in popularity recently as reagents for electrophilic fluorination; however, these are expensive, produce stoichiometric amine by-products and are ultimately derived from elemental fluorine with all the challenges in handling this poses. Of great interest therefore would be electrophilic fluorination protocols in which the source of fluorine is derived from the safer and more easily handled fluoride ion. One approach to achieve this uses I(III)-based electrophilic fluorinating agents, which can be produced by ligand exchange with fluoride at a suitable I(III)-partner (Scheme 17.2).[4] Pattison *et. al.* have shown that ligand exchange

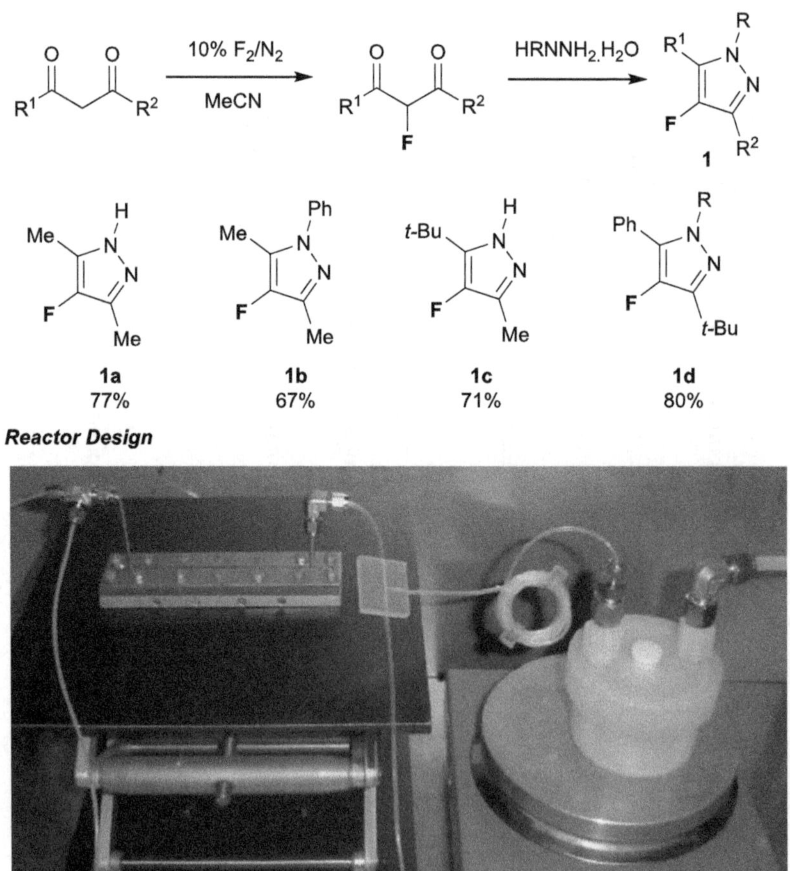

Scheme 17.1 Elemental fluorination using flow microreactors to give pyrazole derivatives (image taken from ref. 3).

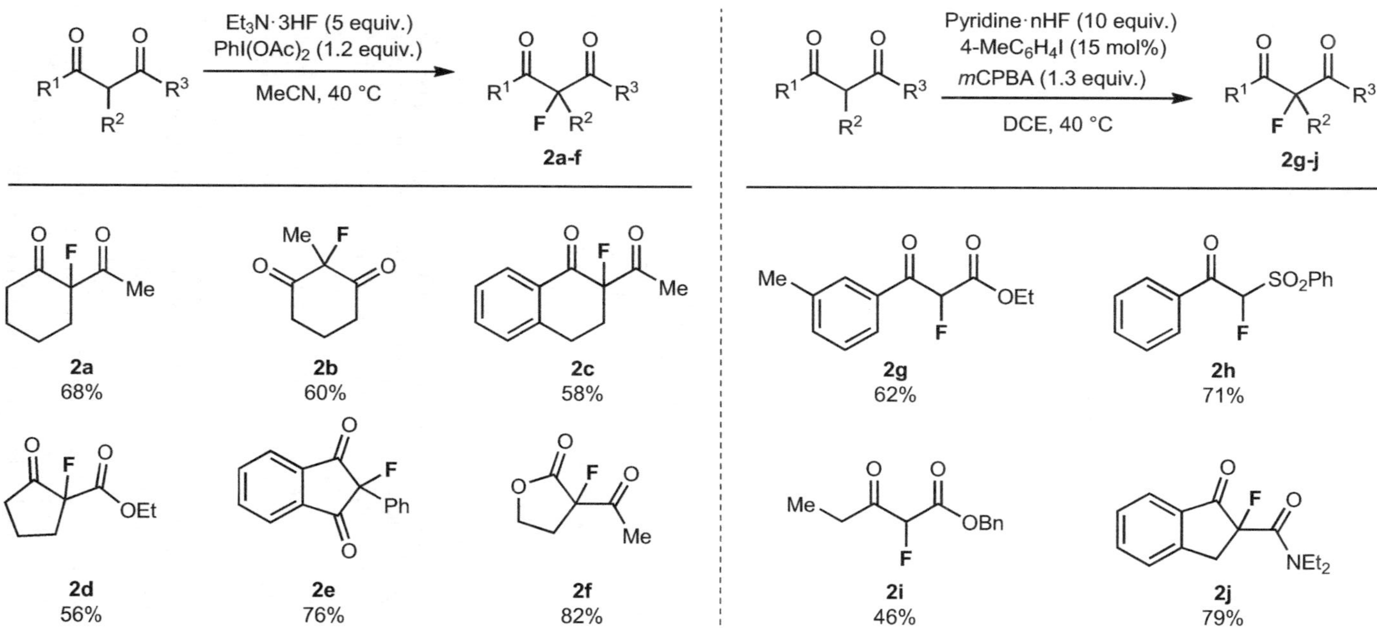

Scheme 17.2 Electrophilic fluorination mediated by fluoride exchange at I(III).

of acetate for fluoride occurs at $PhI(OAc)_2$ using $Et_3N \cdot 3HF$ to yield the active electrophilic fluorinating agent $PhIF_2$ *in situ*.[5] This mixture was shown to be an efficient reagent for the electrophilic fluorination of various 1,3-dicarbonyl compounds **2**. Of course, this approach yields PhI as a stoichiometric by-product. Kita and Shibata demonstrated that $PhIF_2$ for carbonyl fluorination and alkene aminofluorination can be generated *in situ* from an iodoarene in the presence of pyridine · nHF using *m*cpba as a stoichiometric oxidant.[6]

17.3 Nucleophilic Fluorination

Nucleophilic fluorination processes involving fluoride ion salts are in many ways much more attractive than electrophilic processes in terms of safety, ease of handling and cost. These fluoride salts are mass-produced in a sustainable fashion by reaction of metal salts such as carbonates or hydroxides with HF. They typically react in nucleophilic displacement reactions, meaning the only byproduct will be an easily removed metal salt. However, nucleophilic fluorination chemistry is much less well established than electrophilic fluorination chemistry. Many fluoride salts are very hygroscopic, and on complexation with water become solvated and poorly nucleophilic. In unsolvated form fluoride is also a strong base and this can lead to undesired side reactions, such as eliminations. Reagents such as tetrabutylammonium difluorotriphenylsilicate (TBAT) have been developed to provide anhydrous, more nucleophilic sources of fluorine, but these give stoichiometric quantities of silicon-containing waste.

A 'greener' solution to this problem has been the discovery that tetrabutylammonium fluoride can form stable complexes with *tert*-butanol (TBAF · 4*t*-BuOH).[7] This complex has low hygroscopicity and basicity, but good nucleophilicity, and can be stored in air without significant water uptake. This reagent was used in the fluorination of various terminal bromides and tosylates by nucleophilic displacement, to give the terminal fluoride **3** in good yield. This approach showed significantly fewer by-products due to elimination **4** or hydroxylation than the use of either standard or anhydrous TBAF, making it an attractive reagent for nucleophilic fluorination by S_N2 displacement (Scheme 17.3). The *tert*-butanol can be readily removed and disposed of after an aqueous wash; however, chromatography is often required to obtain pure products.

Whilst S_N2 displacement is a viable strategy for aliphatic fluorination, nucleophilic fluorination of aromatic systems is much more challenging. The traditional approach to aromatic fluorination is the Balz–Schiemann reaction, which is the thermal decomposition of aryl diazonium tetrafluoroborate salts, which are potentially explosive reagents. This decomposition also gives the toxic, fuming gas BF_3 as a by-product and this should be removed in a basic aqueous scrubber. Modern alternatives should seek to provide safer and cleaner aromatic fluorination protocols.

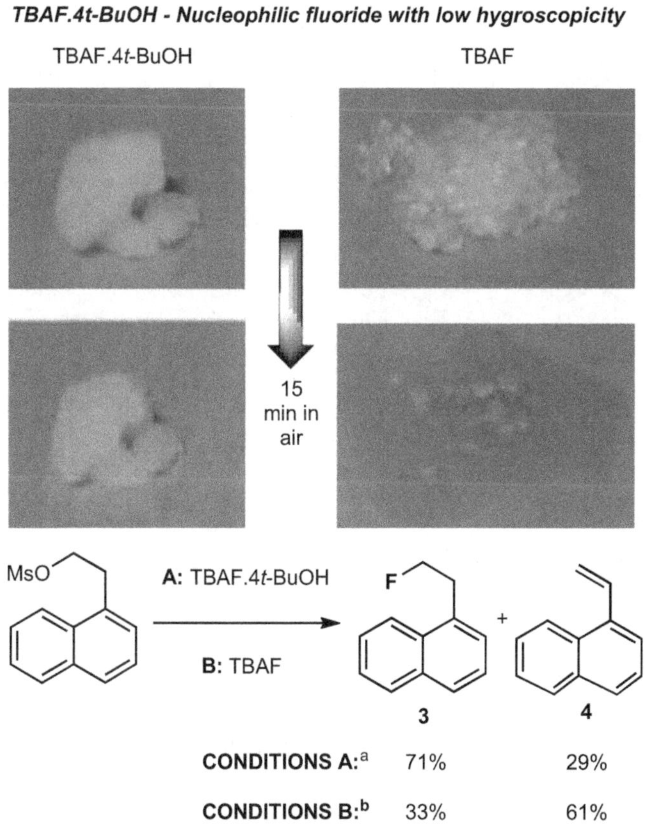

Scheme 17.3 Nucleophilic fluorination using TBAF·4*t*-BuOH. [a]Conditions A: reactions were carried out with 2.0 equiv. of TBAF(t-BuOH)$_4$ at 70 °C for 1 h in CH_3CN. [b]Conditions B: reactions were carried out with 2.0 equiv. of TBAF at 70 °C for 1 h in CH_3CN.
Adapted from ref. 7 with permission from John Wiley and Sons. Copyright © 2008 Wiley-VCH Verlag GmbH & Co. KGaA, Weinheim.

An attractive green approach for the synthesis of fluoroarenes would be a metal-catalysed cross-coupling process of an aryl halide or triflate with fluoride ions, which would be highly atom-economic. Palladium has proven to be the ideal catalyst in a range of carbon–carbon and carbon–heteroatom coupling reactions; however, the development of such a coupling reaction for fluorination has proven to be challenging. Palladium forms extremely strong bonds to fluorine, meaning that carbon–fluorine reductive elimination to release a fluoroarene product is a high-energy process and difficult to achieve.

Buchwald and co-workers successfully developed a catalytic fluorination of aryl triflates in 2009 using [(cinnamyl)PdCl]$_2$ as the catalyst, *t*-BuBrettPhos (**L1**), a bulky biarylphosphine ligand, and caesium fluoride as the fluoride salt.[8] The success of this coupling relied on the formation of a reactive monomeric 14-electron intermediate **5**, which was susceptible to reductive

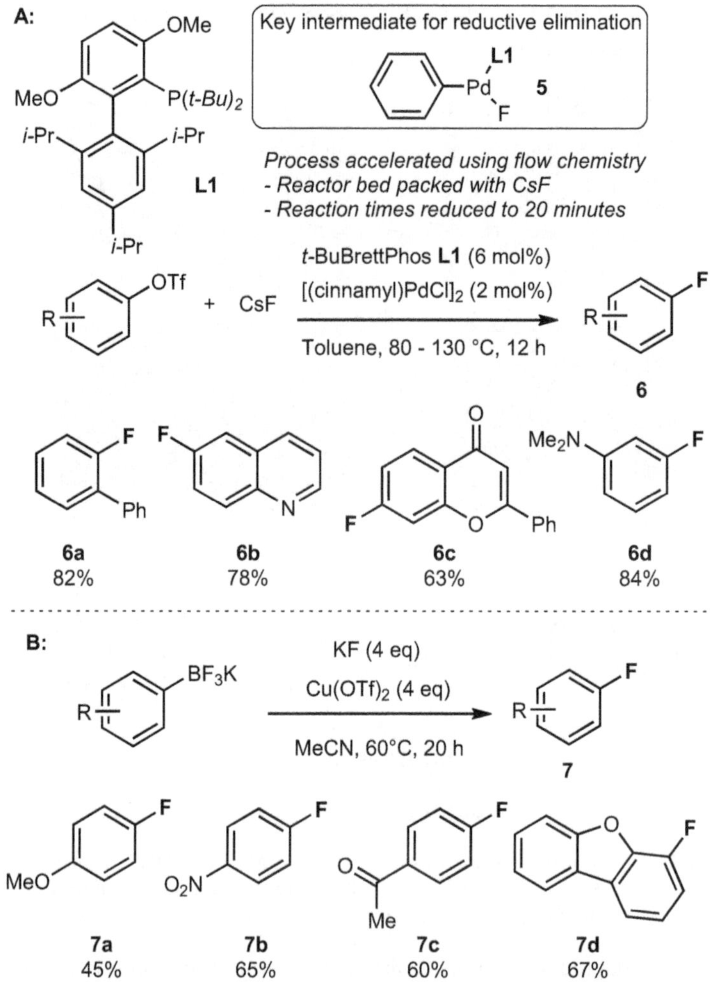

Scheme 17.4 (A) Palladium-catalysed nucleophilic fluorination of aryl triflates. (B) Copper-mediated fluorination of aryl potassium trifluoroborates using potassium fluoride.

elimination of fluoroarene **6** on thermolysis. This process was suitable for the fluorination of a range of electron-rich and electron-poor aryl and heteroaryl triflates (Scheme 17.4A), although fairly forcing conditions were required in some cases. The process could be accelerated by using flow chemistry; a reactor was designed, packed with a bed of caesium fluoride and then a mixture of catalyst and substrate pumped through at high temperature (120 °C).[9] This flow strategy was successful in reducing reaction times from 12 hours to 20 minutes, mainly due to the large excess of fluoride that could be introduced using the packed-bed reactor approach and also because the flow reactor allowed the solvent (toluene) to be safely superheated above its boiling point using a back-pressure regulator.

Improvements to the sustainability of this approach may include supporting the catalyst on the packed-bed reactor meaning it can be easily reused and also simplifying purification.

In terms of cost and environmental impact, potassium fluoride could be considered to be the ideal fluoride source. It is prepared on a multi-tonne scale by reaction of potassium carbonate with HF and is the second most commonly utilised source of fluoride ions in industrial processes after HF, but obviously has many advantages in terms of safety and handling. However, it suffers particularly from high hygroscopicity and low solubility in organic solvents, meaning it is utilised much less often than would be desired for fluorination reactions. However, examples are emerging of its use in aromatic fluorination. Recently, Sanford reported an efficient fluorination of aryl trifluoroborate salts using potassium fluoride, mediated by copper(II) triflate (Scheme 17.4B).[10] The $Cu(OTf)_2$ undergoes disproportionation to form a Cu(III) intermediate that more readily undergoes reductive elimination of a fluoroarene (7). However, before such a process could truly be considered 'green', the large excess of copper present in this reaction should be reduced to a sub-stoichiometric amount for a catalytic process, and oxidation to Cu(III) should ideally be promoted by a benign oxidant, such as oxygen or air.

17.4 C–H Fluorination

The majority of fluorination procedures rely on the substitution of a functional group, for example another halogen atom, which is then lost as waste in the substitution process, and must be responsibly disposed of. A valuable improvement in terms of green chemistry would be processes that allow direct replacement of C–H bonds, leading to the formation of clean by-products such as water. Many challenges exist in the development of such chemistry, not least the high strength of C–H bonds and the ubiquitous nature of C–H bonds in organic molecules, meaning the selective activation of a single C–H bond is challenging. To circumvent these problems, many C–H activation processes use directing groups that promote metal-catalysed cleavage of a single C–H bond selectively. However, this directing group approach may limit the range of molecules that can be made, and the directing groups may be difficult to remove.

C–H activation reactions are beginning to be applied to fluorination processes. This work was pioneered by Sanford and Yu, who independently showed that C–H fluorination could be mediated by palladium catalysts in the presence of electrophilic fluorine sources (Scheme 17.5A). Sanford used heterocyclic rings to direct catalytic fluorination on an adjacent benzene ring or benzylic alkyl group,[11] whilst Yu used triflamide directing groups to direct *ortho* fluorination.[12] The directing groups lead to the activation of a C–H bond through a cyclometallation process. The key to this fluorination chemistry was the formation of intermediates based on high oxidation state Pd, likely to be Pd(IV), which are high-energy intermediates with a reduced

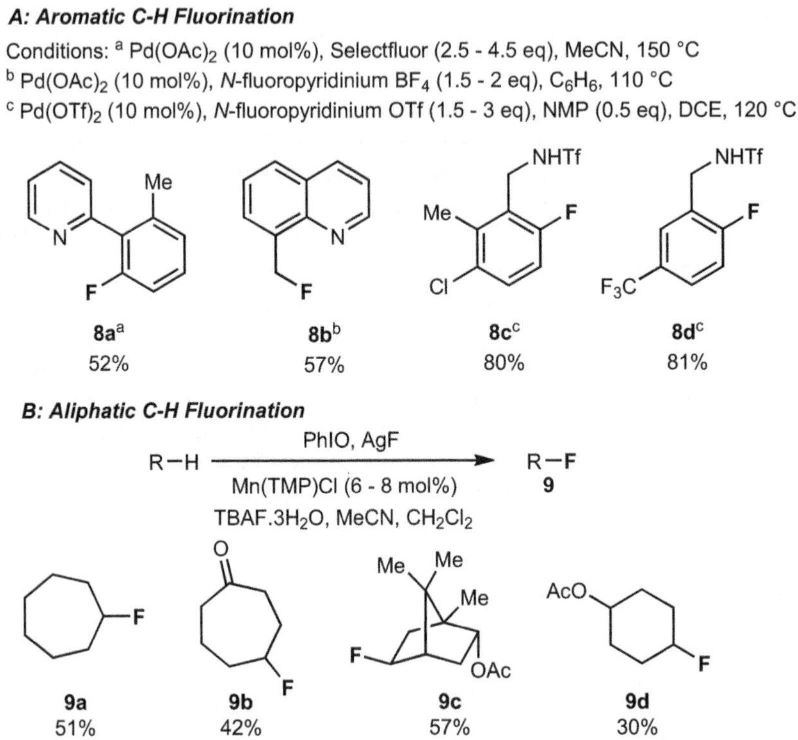

Scheme 17.5 Catalytic C–H fluorination processes.

energy barrier towards C–F reductive elimination. Electrophilic sources of fluorine such as Selectfluor and *N*-fluoropyridinium salts were used which are highly oxidising, and compatible with the oxidation of Pd(II) to Pd(IV).

C–H fluorination without directing groups is also of great interest, and has been achieved recently in a range of aliphatic systems. Groves has used manganese(III) porphyrin complexes to promote C–H bond fluorination at a broad range of secondary alkyl carbons under oxidising conditions (Scheme 17.5B).[13,14] Mn(III) is oxidised to a Mn(V)-oxo complex by the hypervalent iodine(III) oxidant PhIO. This metal-oxo complex then performs a radical-based C–H bond cleavage to form an alkyl radical, which is then fluorinated. This approach was used to activate fluorinate single C–H bonds in complex substrates selectively.

Although C–H bond activation is an extremely promising strategy for the development of sustainable chemical processes, by minimising waste and extra reaction steps needed to introduce functional handles, the above examples do require further work before they can truly be described as 'green'. For example, some of the harsh stoichiometric oxidants required should be replaced by oxygen or air to improve the environmental impact of these reactions.

17.5 Trifluoromethylation

Traditional approaches towards the formation of trifluoromethyl groups can involve harsh reagents and conditions and be incompatible with a 'green' approach. For example, carboxylic acids can be converted to trifluoromethyl groups by reaction with sulfur tetrafluoride, whilst trichloromethyl groups can be perfluorinated using antimony trifluoride. Both of these are reagents with significant hazards and environmental effects. Even milder reagents for the introduction of trifluoromethyl functionality, such as Me_3SiCF_3, are expensive and produce stoichiometric quantities of silicon-containing waste.

Chemists are therefore extremely keen to establish mild reagents that will allow the facile introduction of trifluoromethyl groups into organic molecules. Possibly the ideal reagent for this is fluoroform (CHF_3), several thousand tonnes of which are produced annually as a by-product from the fluorination of chloroform to $CHClF_2$ with HF during Teflon production. Fluoroform is a very potent greenhouse gas, whose warming effect is more than 10 000 times greater than CO_2. Although there is currently no legal requirement in the EU to destroy fluoroform, much is destroyed by incineration at 1473 K, yielding water, carbon dioxide and HF, which need to be removed by scrubbing processes. To utilise this by-product in useful trifluoromethylation reactions would be much more satisfactory. Fluoroform currently sees little application in trifluoromethylation as it is a relatively unreactive gaseous reagent that is fairly unstable when deprotonated to form CF_3^-.

In 2012, Prakash discovered that fluoroform could be deprotonated using KHMDS at low temperatures (ranging from −78 °C to −5 °C) and the resulting anion could be trapped *in situ* using various electrophiles including trialkylsilyl chlorides, borates and aldehydes and ketones (Scheme 17.6A).[15] This was the first truly general synthetic application of fluoroform as a trifluoromethylating agent. Grushin has also developed the synthesis of a $CuCF_3$ species from fluoroform using KO*t*-Bu and CuCl. This species could be stabilised by the addition of $Et_3N \cdot 3HF$ and was shown to participate in coupling reactions with iodo- and bromo-arenes,[16] as well as with boronic acids.[17]

Grushin also very recently demonstrated that fluoroform can undergo C–H activation in the presence of [(dppp)Pd(Ph)(OH)] and *n*-Bu_3P as a Lewis basic promoter (Scheme 17.6B).[18] Although this has not yet led to catalytic trifluoromethylation reactions using fluoroform, it is a significant advance towards this important goal for sustainable trifluoromethylation chemistry.

Other inexpensive, easy-to-handle trifluoromethyl sources are also of great interest. For example, trifluoroacetate salts are potential trifluoromethylating agents through decarboxylation, the only by-product being CO_2. These salts are derived from trifluoroacetic acid, which is produced on an industrial scale by electrochemical fluorination of acetyl chloride with HF. After pioneering work by Chambers[19] and Kondo[20] in the 1980s on the trifluoromethylation of iodoarenes mediated by Cu, Duan has reported the

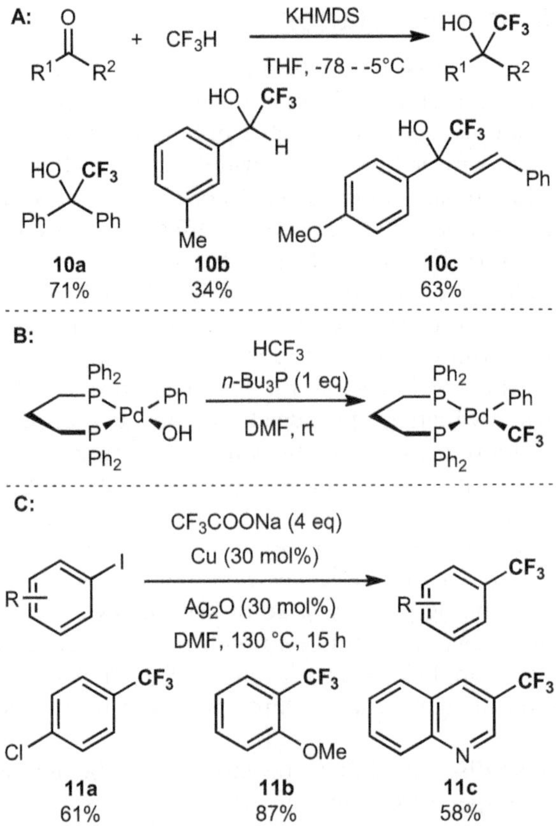

Scheme 17.6 (A) KHMDS-mediated trifluoromethylation with fluoroform. (B) Pd-mediated C–H activation of fluoroform. (C) Decarboxylative trifluoromethylation with sodium trifluoroacetate.

copper metal-catalysed decarboxylative trifluoromethylation of iodoarenes, in which the decarboxylation was promoted by the addition of 30 mol% of silver(I) oxide (Scheme 17.6C).[21]

17.6 Other Halogenation Processes

Bromine and chlorine are both hazardous, corrosive and toxic substances whose handling should ideally be avoided. Safer alternatives, such as *N*-halosuccinimides, have been developed to replace these elemental halogens in electrophilic halogenation processes. However, the atom economy of these reagents is poor and they produce stoichiometric quantities of amide by-products, which must be removed. Again, simple halide salts, such as potassium chloride or bromide, are ideal halogenating agents due to their inert nature and easy handling. Strategies to engage these nucleophilic salts in electrophilic halogenation processes are therefore of immense value.

The redox potentials of chloride and bromide are significantly lower than that of fluoride, and it is possible to oxidise these heavier halides to the corresponding elemental halogen by chemical means. If clean chemical oxidants, such as hydrogen peroxide or oxygen, can be used for this oxidation then such chemistry will be of great interest in the development of 'green' halogenation processes (indeed this approach is used for halogenation by biological systems).

Vanadium and molybdenum complexes in high oxidation states such as V(v) and Mo(vi) have been used as catalysts for oxidative halogenation reactions using halide salts and hydrogen peroxide as the oxidising agent. This approach has proved effective for the halogenation of aromatic rings and alkenes. In certain cases, particularly on a laboratory scale, no catalyst may be required for effective oxidative halogenation processes, although these may require more forcing conditions and higher temperatures than catalysed processes. The likely active halogenating agent in these reactions is the hypohalous acid (HOX), and these are especially attractive 'green' halogenations as the only by-product produced is water. Again, uncatalysed oxidative halogenation processes have been applied using hydrogen peroxide as the oxidant for the chlorination and bromination of aromatics, alkenes and carbonyl compounds (Scheme 17.7A).[22]

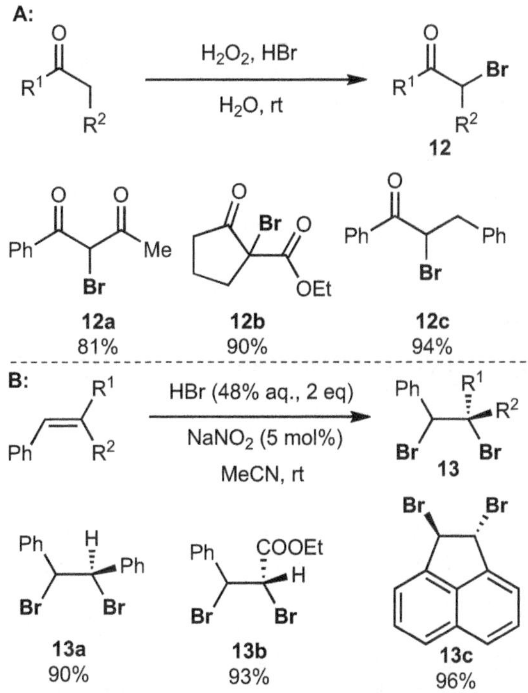

Scheme 17.7 (A) Oxidative bromination of carbonyl compounds with hydrogen peroxide and HBr. (B) Oxidative aerobic bromination of alkenes.

The replacement of hydrogen peroxide with oxygen or air for such sustainable oxidative halogenation processes is desirable in terms of safety. The bromination of alkenes has been achieved using HBr and 5 mol% $NaNO_2$ in air (Scheme 17.7B).[23] Under the aerobic, acidic reaction conditions, $NaNO_2$ forms species such as NOBr, which decompose to form Br_2, which mediates the bromination reaction. These oxidative halogenation approaches minimise the quantities of dangerous reactive elemental halogens in a reaction and ensure the only reagents that must be directly handled are the safer halide salts or acids.

17.7 Summary

- This short chapter has given highlights of some of the recent work in 'green' and sustainable halogenation reactions.
- Chemists are now starting to demonstrate that we have viable alternatives to the use of toxic and dangerous elemental halogens, or that when these must be used we now have strategies to handle them in a safe and environmentally responsible way.
- Our use of safe and easily handled halide salts as halogenating agents is now increasing and approaches are being developed to use these nucleophilic reagents in electrophilic oxidative halogenation processes.
- However, challenges still remain in the mild and selective halogenation of a range of systems, and it is likely that the growing use of metal-catalysed reactions in this area will lead to the development of a valuable range of new 'green' halogenation reactions.
- The widespread appearance of the halogens in pharmaceuticals and synthetic intermediates means it is essential that more effort is made towards our ability to perform halogenation reactions conveniently and with minimal environmental impact.

Acknowledgements

GP is funded by an IAS Global Research Fellowship at the University of Warwick.

References

1. D. O'Hagan, *Chem. Soc. Rev.*, 2008, **37**, 308–319.
2. S. Purser, P. R. Moore, S. Swallow and V. Gouverneur, *Chem. Soc. Rev.*, 2008, **37**, 320–330.
3. J. R. Breen, G. Sandford, D. S. Yufit, J. A. K. Howard, J. Fray and B. Patel, *Beilstein J. Org. Chem.*, 2011, **7**, 1048–1054.
4. G. C. Geary, E. G. Hope, K. Singh and A. M. Stuart, *Chem. Commun.*, 2013, **49**, 9263–9265.
5. T. J. Nash and G. Pattison, *Eur. J. Org. Chem.*, 2015, **2015**, 3779–3786.

6. S. Suzuki, T. Kamo, K. Fukushi, T. Hiramatsu, E. Tokunaga, T. Dohi, Y. Kita and N. Shibata, *Chem. Sci.*, 2014, **5**, 2754–2760.
7. D. W. Kim, H.-J. Jeong, S. T. Lim and M.-H. Sohn, *Angew. Chem., Int. Ed.*, 2008, **47**, 8404–8406.
8. D. A. Watson, M. Su, G. Teverovskiy, Y. Zhang, J. García-Fortanet, T. Kinzel and S. L. Buchwald, *Science*, 2009, **325**, 1661–1664.
9. T. Noël, T. J. Maimone and S. L. Buchwald, *Angew. Chem., Int. Ed.*, 2011, **50**, 8900–8903.
10. Y. Ye, S. D. Schimler, P. S. Hanley and M. S. Sanford, *J. Am. Chem. Soc.*, 2013, **135**, 16292–16295.
11. K. L. Hull, W. Q. Anani and M. S. Sanford, *J. Am. Chem. Soc.*, 2006, **128**, 7134–7135.
12. X. Wang, T.-S. Mei and J.-Q. Yu, *J. Am. Chem. Soc.*, 2009, **131**, 7520–7521.
13. W. Liu, X. Huang, M.-J. Cheng, R. J. Nielsen, W. A. Goddard and J. T. Groves, *Science*, 2012, **337**, 1322–1325.
14. W. Liu and J. T. Groves, *Angew. Chem., Int. Ed.*, 2013, **52**, 6024–6027.
15. G. K. S. Prakash, P. V. Jog, P. T. D. Batamack and G. A. Olah, *Science*, 2012, **338**, 1324–1327.
16. A. Zanardi, M. A. Novikov, E. Martin, J. Benet-Buchholz and V. V. Grushin, *J. Am. Chem. Soc.*, 2011, **133**, 20901–20913.
17. P. Novák, A. Lishchynskyi and V. V. Grushin, *Angew. Chem., Int. Ed.*, 2012, **51**, 7767–7770.
18. S. Takemoto and V. V. Grushin, *J. Am. Chem. Soc.*, 2013, **135**, 16837–16840.
19. G. E. Carr, R. D. Chambers, T. F. Holmes and D. G. Parker, *J. Chem. Soc. Perkin Trans. 1*, 1988, 921–926.
20. K. Matsui, E. Tobita, M. Ando and K. Kondo, *Chem. Lett.*, 1981, **10**, 1719–1720.
21. Y. Li, T. Chen, H. Wang, R. Zhang, K. Jin, X. Wang and C. Duan, *Synlett*, 2011, **2011**, 1713–1716.
22. A. Podgorsek, S. Stavber, M. Zupan and J. Iskra, *Green Chem.*, 2007, **9**, 1212–1218.
23. A. Podgorsek, M. Eissen, J. Fleckenstein, S. Stavber, M. Zupan and J. Iskra, *Green Chem.*, 2009, **11**, 120–126.

Subject Index